Lecture Notes in Mathematics

Jean-Pierre Serre

Lie Algebras and Lie Groups

1964 Lectures given at Harvard University

Springer-Verlag
Berlin Heidelberg New York
London Paris Tokyo
Hong Kong Barcelona
Budapest

Author

Jean-Pierre Serre
Collège de France
3, rue d'Ulm
75005 Paris, France

Mathematics Subject Classification (1991): 17B

2nd edition
originally published by: W. A. Benjamin, Inc., New York, 1965

ISBN 3-540-55008-9 Springer-Verlag Berlin Heidelberg New York
ISBN 0-387-55008-9 Springer-Verlag New York Berlin Heidelberg

Printing and binding: Druckhaus Beltz, Hemsbach/Bergstr.
46/3140-543210 - Printed on acid-free paper

Contents

Part I – Lie Algebras

Introduction

The main general theorems on Lie Algebras are covered, roughly the content of Bourbaki's Chapter I.

I have added some results on free Lie algebras, which are useful, both for Lie's theory itself (Campbell-Hausdorff formula) and for applications to pro-p-groups.

Lack of time prevented me from including the more precise theory of semisimple Lie algebras (roots, weights, etc.); but, at least, I have given, as a last Chapter, the typical case of $\mathfrak{sl}_n$.

This part has been written with the help of F. Raggi and J. Tate. I want to thank them, and also Sue Golan, who did the typing for both parts.

Jean-Pierre Serre

Harvard, Fall 1964

Chapter I. Lie Algebras: Definition and Examples

Let k be a commutative ring with unit element, and let A be a k-module, then A is said to be a k-algebra if there is given a k-bilinear map $A \times A \to A$ (i.e., a k-homomorphism $A \otimes_k A \to A$).

As usual we may define left, right and two-sided ideals and therefore quotients.

Definition 1. A *Lie algebra* over k is an algebra with the following properties:

1). The map $A \otimes_k A \to A$ admits a factorization

$$A \otimes_k A \to \bigwedge^2 A \to A$$

i.e., if we denote the image of (x, y) under this map by $[x, y]$ then the condition becomes

$$[x, x] = 0 \qquad \text{for all } x \in k.$$

2). $[[x, y], z] + [[y, z], x] + [[z, x], y] = 0$ (Jacobi's identity)

The condition 1) implies $[x, y] = -[y, x]$.

Examples. (i) Let k be a complete field with respect to an absolute value, let G be an analytic group over k, and let $\mathfrak{g}$ be the set of tangent vectors to G at the origin. There is a natural structure of Lie algebra on $\mathfrak{g}$.

(For an algebraic analogue of this, see example (v) below.)

(ii) Let $\mathfrak{g}$ be any k-module. Define $[x, y] = 0$ for all $x, y \in \mathfrak{g}$. Such a $\mathfrak{g}$ is called a *commutative* Lie algebra.

(ii′) If in the preceding example we take $\mathfrak{g} \oplus \bigwedge^2 \mathfrak{g}$ and define

$$[x, y] = x \wedge y$$
$$[x, y \wedge z] = 0$$
$$[x \wedge y, z] = 0$$
$$[x \wedge y, z \wedge t] = 0$$

for all $x, y, z, t \in \mathfrak{g}$, then $\mathfrak{g} \oplus \bigwedge^2 \mathfrak{g}$ is a Lie algebra.

(iii) Let A be an associative algebra over k and define $[x, y] = xy - yx$, $x, y \in A$. Clearly A with this product satisfies the axioms 1) and 2).

Definition 2. Let A be an algebra over k. A *derivation* $D : A \to A$ is a k-linear map with the property $D(x \cdot y) = Dx \cdot y + x \cdot Dy$.

(iv) The set $\mathrm{Der}(A)$ of all derivations of an algebra A is a Lie algebra with the product $[D, D'] = DD' - D'D$.

We prove it by computation:

$$\begin{aligned}
[D,D'](x\cdot y) &= DD'(x\cdot y) - D'D(x\cdot y)\\
&= D(D'x\cdot y + x\cdot D'y) - D'(Dx\cdot y + x\cdot Dy)\\
&= DD'x\cdot y + D'x\cdot Dy + Dx\cdot D'y + x\cdot DD'y\\
&\quad - D'Dx\cdot y - Dx\cdot D'y - D'x\cdot Dy - x\cdot D'Dy\\
&= DD'x\cdot y + x\cdot DD'y - D'Dx\cdot y - x\cdot D'Dy\\
&= [D,D']x\cdot y + x\cdot [D,D']y\ .
\end{aligned}$$

Theorem 3. *Let* $\mathfrak{g}$ *be a Lie algebra. For any* $x \in \mathfrak{g}$ *define a map* $\operatorname{ad} x : \mathfrak{g} \to \mathfrak{g}$ *by* $\operatorname{ad} x(y) = [x,y]$, *then:*

1) $\operatorname{ad} x$ *is a derivation of* $\mathfrak{g}$.

2) *The map* $x \mapsto \operatorname{ad} x$ *is a Lie homomorphism of* $\mathfrak{g}$ *into* $\operatorname{Der}(\mathfrak{g})$.

Proof.

$$\begin{aligned}
\operatorname{ad} x[y,z] &= [x,[y,z]]\\
&= -[y,[z,x]] - [z,[x,y]]\\
&= [[x,y],z] + [y,[x,z]]\\
&= [\operatorname{ad} x(y), z] + [y, \operatorname{ad} x(z)]\ ,
\end{aligned}$$

hence, 1) is equivalent to the Jacobi identity. Now

$$\begin{aligned}
\operatorname{ad}[x,y](z) &= [[x,y],z]\\
&= -[[y,z],x] - [[z,x],y]\\
&= [x,[y,z]] - [y,[x,z]]\\
&= \operatorname{ad} x \operatorname{ad} y(z) - \operatorname{ad} y \operatorname{ad} x(z)\\
&= [\operatorname{ad} x, \operatorname{ad} y](z)\ ,
\end{aligned}$$

hence 2) is also equivalent to the Jacobi identity.

(v) *The Lie algebra of an algebraic matrix group.*

Let k be a commutative ring and let $A = M_n(k)$ be the algebra of $n \times n$-matrices over k.

Given a set of polynomials $P_\alpha(X_{ij})$, $1 \le i,j \le n$, a zero of (P_α) is a matrix $x = (x_{ij})$ such that $x_{ij} \in k$, $P_\alpha(x_{ij}) = 0$ for all α.

Let $G(k)$ denote the set of zeroes of (P_α) in $A^* = \mathrm{GL}_n(k)$. If k' is any associative, commutative k-algebra we have analogously $G(k') \subset M_n(k')$.

Definition 4. The set (P_α) defines an algebraic group over k if $G(k')$ is a subgroup of $\mathrm{GL}_n(k')$ for all associative, commutative k-algebras k'.

The orthogonal group is an example of an algebraic group (equation: ${}^tX \cdot X = 1$, where ${}^tX$ denotes the transpose of X).

Now, let k' be the k-algebra which is free over k with basis $\{1, \varepsilon\}$ where $\varepsilon^2 = 0$, i.e., $k' = k[\varepsilon]$.

Theorem 5. *Let* $\mathfrak{g}$ *be the set of matrices* $X \in M_n(k)$ *such that*

$$1 + \varepsilon X \in G(k[\varepsilon]) .$$

Then $\mathfrak{g}$ *is a Lie subalgebra of* $M_n(k)$.

We have to prove that $X, Y \in \mathfrak{g}$ implies $\lambda X + \mu Y \in \mathfrak{g}$, if $\lambda, \mu \in k$ and $XY - YX \in \mathfrak{g}$.

To prove that, note first that

$$P_\alpha(1 + \varepsilon X) = 0 \text{ for all } \alpha \quad \Longleftrightarrow \quad X \in \mathfrak{g}$$

and, since $\varepsilon^2 = 0$, we have

$$P_\alpha(1 + \varepsilon X) = P_\alpha(1) + dP_\alpha(1)\varepsilon X .$$

But $1 \in G(k)$, i.e. $P_\alpha(1) = 0$; therefore

$$P_\alpha(1 + \varepsilon X) = dP_\alpha(1)\varepsilon X .$$

Hence, $\mathfrak{g}$ is a submodule of $M_n(k)$.

We introduce now an auxiliary algebra k'' given by $k'' = k[\varepsilon, \varepsilon', \varepsilon\varepsilon']$ where $\varepsilon^2 = \varepsilon'^2 = 0$ and $\varepsilon'\varepsilon = \varepsilon\varepsilon'$, i.e., $k'' = k[\varepsilon] \otimes_k k[\varepsilon']$.

Let $X, Y \in \mathfrak{g}$, so we have

$$g = 1 + \varepsilon X \in G(k[\varepsilon]) \subset G(k'')$$
$$g' = 1 + \varepsilon' Y \in G(k[\varepsilon']) \subset G(k'')$$

$gg' = (1 + \varepsilon X)(1 + \varepsilon' Y) = 1 + \varepsilon X + \varepsilon' Y + \varepsilon\varepsilon' XY$
$g'g = 1 + \varepsilon X + \varepsilon' Y + \varepsilon\varepsilon' YX$.
Write $Z = [X, Y]$; we have

$$gg' = g'g(1 + \varepsilon\varepsilon' Z) .$$

Since $gg', g'g \in G(k'')$, it follows that

$$1 + \varepsilon\varepsilon' Z \in G(k'') .$$

But the subalgebra $k[\varepsilon\varepsilon']$ of k'' may be identified with $k[\varepsilon]$. It then follows that $1 + \varepsilon Z \in G(k[\varepsilon])$, hence $Z \in \mathfrak{g}$, q.e.d.

Example. The Lie algebra of the orthogonal group is the set of matrices X such that $(1 + \varepsilon X)(1 + \varepsilon({}^tX)) = 1$, i.e., $X + {}^tX = 0$.

(vi) *Construction of Lie algebras from known ones.*

a) Let $\mathfrak{g}$ be a Lie algebra and let $J \subset \mathfrak{g}$ an ideal, then $\mathfrak{g}/J$ is a Lie algebra.

b) Let $(\mathfrak{g}_i)_{i \in I}$ be a family of Lie algebras, then $\prod_{i \in I} \mathfrak{g}_i$ is a Lie algebra.

c) Suppose $\mathfrak{g}$ is a Lie algebra, $\mathfrak{a} \subset \mathfrak{g}$ is an ideal and $\mathfrak{b}$ is a subalgebra of $\mathfrak{g}$, then $\mathfrak{g}$ is called a *semidirect product* of $\mathfrak{b}$ by $\mathfrak{a}$ if the natural map $\mathfrak{g} \to \mathfrak{g}/\mathfrak{a}$

induces an isomorphism $\mathfrak{b} \xrightarrow{\sim} \mathfrak{g}/\mathfrak{a}$. If so, and if $x \in \mathfrak{b}$, then $\operatorname{ad} x$ maps $\mathfrak{a}$ into $\mathfrak{a}$ so that $\operatorname{ad}_{\mathfrak{a}} x \in \operatorname{Der}(\mathfrak{a})$, i.e., we have a Lie homomorphism $\theta : \mathfrak{b} \to \operatorname{Der}(\mathfrak{a})$.

Theorem 6. *The structure of $\mathfrak{g}$ is determined by $\mathfrak{a}$, $\mathfrak{b}$ and θ, and these can be given arbitrarily.*

Proof. Since $\mathfrak{g}$ is the direct sum of $\mathfrak{a}$ and $\mathfrak{b}$ as a k-module and since multiplication is bilinear and anticommutative we have to consider the product $[x, y]$ in the following three cases:

$$x, y \in \mathfrak{a}$$
$$x, y \in \mathfrak{b}$$
$$x \in \mathfrak{b},\ y \in \mathfrak{a}\ .$$

In the first case $[x, y]$ is given in $\mathfrak{a}$, in the second one $[x, y]$ is given in $\mathfrak{b}$ and in the last one we have

$$[x, y] = \operatorname{ad} x(y) = \theta(x)y\ .$$

Conversely, given the Lie algebras $\mathfrak{a}$ and $\mathfrak{b}$ and a Lie homomorphism

$$\theta : \mathfrak{b} \to \operatorname{Der}(\mathfrak{a})\ ,$$

we can construct a Lie algebra $\mathfrak{g}$ which is a semidirect product of $\mathfrak{b}$ by $\mathfrak{a}$ in such a way that $\theta(x) = \operatorname{ad}_{\mathfrak{a}} x$, where $\operatorname{ad}_{\mathfrak{a}} x$ is the restriction to $\mathfrak{a}$ of $\operatorname{ad}_{\mathfrak{g}} x$, for $x \in \mathfrak{b}$. One has to check that the Jacobi's identity

$$J(x, y, z) = [x, [y, z]] + [y, [z, x]] + [z, [x, y]] = 0$$

holds. There are essentially four cases to be considered:

(a) $x, y, z \in \mathfrak{a}$ – then $J(x, y, z) = 0$ because $\mathfrak{a}$ is a Lie algebra.

(b) $x, y \in \mathfrak{a}$, $z \in \mathfrak{b}$ – $J(x, y, z) = 0 \iff \theta(z)$ is a derivation of $\mathfrak{a}$.

(c) $x \in \mathfrak{a}$, $y, z \in \mathfrak{b}$ – $J(x, y, z) = 0 \iff \theta([y, z]) = \theta(y)\theta(z) - \theta(z)\theta(y)$.

(d) $x, y, z \in \mathfrak{b}$ – $J(x, y, z) = 0$ because $\mathfrak{b}$ is a Lie algebra.

Chapter II. Filtered Groups and Lie Algebras

1. Formulae on commutators

Let G be a group and let $x, y, z \in G$. We will use the following notations:

(i) $x^y = y^{-1}xy$, hence the map $G \to G$ given by $x \mapsto x^y$ is an automorphism of G, and we have the relation $(x^y)^z = x^{yz}$.

(ii) $(x, y) = x^{-1}y^{-1}xy$ which is called the commutator of x and y.

Proposition 1.1. *We have the identities:*

(1) $xy = yx^y = yx(x,y)$, $x^y = x(x,y)$, $(x,x) = 1$, $(y,x) = (x,y)^{-1}$.

(2) $(x, yz) = (x,z)(x,y)^z$.

(2′) $(xy, z) = (x,z)^y(y,z)$.

(3) $(x^y,(y,z))(y^z,(z,x))(z^x,(x,y)) = 1$.

Proof. (1) is trivial.

(2) From (i) and (1) we have

$$\begin{aligned} x(x,yz) &= x^{yz} \\ &= (x^y)^z \\ &= [x(x,y)]^z \\ &= x^z(x,y)^z = x(x,z)(x,y)^z \end{aligned}$$

and therefore $(x,yz) = (x,z)(x,y)^z$.

(2′)
$$\begin{aligned} xy(xy,z) = (xy)^z &= x^z y^z \\ &= x(x,z)y(y,z) \\ &= xy(x,z)^y(y,z) \end{aligned}$$

and therefore $(xy,z) = (x,z)^y(y,z)$.

(3)
$$\begin{aligned} (x^y,(y,z)) &= y^{-1}x^{-1}yz^{-1}y^{-1}zyy^{-1}xyy^{-1}z^{-1}yz \\ &= y^{-1}x^{-1}yz^{-1}y^{-1}zxz^{-1}yz\ . \end{aligned}$$

Put

$$\begin{aligned} u &= zxz^{-1}yz \\ v &= xyx^{-1}zx \\ w &= yzy^{-1}xy \end{aligned}$$

then $(x^y,(y,z)) = w^{-1}u$.

Analogously (by cyclic permutation)

$$\begin{aligned} (y^z,(z,x)) &= u^{-1}v \\ (z^x,(x,y)) &= v^{-1}w\ . \end{aligned}$$

Hence $(x^y,(y,z))(y^z,(z,x))(z^x,(x,y)) = 1$ q.e.d.

Applications:

Let A, B be subgroups of a group G and let (A,B) denote the subgroup of G generated by the commutators (a,b) for all $a \in A$, $b \in B$.

If A, B, C are normal subgroups of G, then (A,B) is also normal and we have the relation

$$(A,(B,C)) \subset (B,(C,A))(C,(A,B))$$

which follows from 1.1(3).

2. Filtration on a group

Definition 2.1. A *filtration on a group* G is a map $w : G \to \mathbf{R} \cup \{+\infty\}$ satisfying the following axioms:

(1) $w(1) = +\infty$.

(2) $w(x) > 0$ for all $x \in G$.

(3) $w(xy^{-1}) \geq \inf\{w(x), w(y)\}$.

(4) $w((x,y)) \geq w(x) + w(y)$.

It follows from (3) that $w(y^{-1}) = w(y)$. If $\lambda \in \mathbf{R}_+$ we define

$$G_\lambda = \{\, x \in G \mid w(x) \geq \lambda \,\}$$
$$G_\lambda^+ = \{\, x \in G \mid w(x) > \lambda \,\} .$$

The condition (3) shows that G_λ, G_λ^+ are subgroups of G. Moreover, if $x \in G_\lambda$, $y \in G$ then $x^y \equiv x \pmod{G_\lambda^+}$ which follows from the relation

$$w((x,y)) \geq \lambda + w(y) > \lambda .$$

This also proves that G_λ is a normal subgroup of G and since $G_\lambda^+ = \bigcup_{\mu>\lambda} G_\mu$ it follows that G_λ^+ is also a normal subgroup of G.

The family $\{G_\lambda\}$ (resp. $\{G_\lambda^+\}$) is decreasing, i.e., $\lambda < \mu$ implies $G_\lambda \supset G_\mu$ (resp. $G_\lambda^+ \supset G_\mu^+$).

Definition 2.2. For all $\alpha \geq 0$ we define

$$\operatorname{gr}_\alpha G = G_\alpha / G_\alpha^+ \quad \text{and} \quad \operatorname{gr} G = \sum_\alpha \operatorname{gr}_\alpha G .$$

Proposition 2.3.

1) *$\operatorname{gr}_\alpha G$ is an abelian group.*

2) *If $x \in G_\alpha$ let $\bar{x}$ be its image in $\operatorname{gr}_\alpha G$; one has $\overline{(x^y)} = \bar{x}$ for all $y \in G$.*

3) *The map* $c_{\alpha,\beta} : G_\alpha \times G_\beta \to G_{\alpha+\beta}$ *defined by* $x, y \mapsto (x,y)$ *induces a bilinear map* $\bar{c}_{\alpha,\beta} : \mathrm{gr}_\alpha G \times \mathrm{gr}_\beta G \to \mathrm{gr}_{\alpha+\beta} G$.

4) *The maps* $\bar{c}_{\alpha,\beta}$ *can be extended by linearity to* $c : \mathrm{gr}\, G \times \mathrm{gr}\, G \to \mathrm{gr}\, G$ *and this defines a Lie algebra structure on* $\mathrm{gr}\, G$.

Proof. 1) It follows from 2.1(4).

2) It is already proved.

3) Let $x, x' \in G_\alpha$, $y, y' \in G_\beta$, then $(x,y) \in G_{\alpha+\beta}$ and we have to prove that if $u, v \in G_\alpha^+$ then $(xu, y) \equiv (x,y) \bmod G^+_{\alpha+\beta}$, $(x, yv) \equiv (x,y) \bmod G^+_{\alpha+\beta}$. Using 1.1(2′) and (3) we have

$$\begin{aligned}
\overline{(xu,y)} &= \overline{(x,y)^u} + \overline{(u,y)} = \overline{(x,y)} \\
\overline{(x,yv)} &= \overline{(x,v)} + \overline{(x,y)^v} = \overline{(x,y)} \\
\overline{(xx',y)} &= \overline{(x,y)^{x'}} + \overline{(x',y)} = \overline{(x,y)} + \overline{(x',y)} \\
\overline{(x,y'y)} &= \overline{(x,y)} + \overline{(x,y')^y} = \overline{(x,y)} + \overline{(x,y')} \,.
\end{aligned}$$

This proves 3).

4) Let $\xi \in \mathrm{gr}_\alpha G$, $\eta \in \mathrm{gr}_\beta G$ and choose elements $x \in G_\alpha$, $x \in G_\beta$ such that $\bar{x} = \xi$, $\bar{y} = \eta$. Then we have $\overline{(x,y)} = \bar{c}_{\alpha,\beta}(\xi,\eta)$, which we also write $[\xi,\eta]$.

Now if $\xi \in \mathrm{gr}\, G$ then $\xi = \sum_\alpha \xi_\alpha$ where $\xi_\alpha \in \mathrm{gr}_\alpha G$. In order to prove that $[\xi,\xi] = 0$, it is sufficient to prove that $[\xi_\alpha,\xi_\alpha] = 0$ and $[\xi_\alpha,\xi_\beta] = -[\xi_\beta,\xi_\alpha]$. Let $x_\alpha \in G_\alpha$ such that $\bar{x}_\alpha = \xi_\alpha$ for all α. Then we have $[\xi_\alpha,\xi_\alpha] = \overline{(x_\alpha,x_\alpha)} = \bar{1} = 0$, and

$$[\xi_\alpha,\xi_\beta] = \overline{(x_\alpha,x_\beta)} = \overline{(x_\beta,x_\alpha)^{-1}} = -[\xi_\beta,\xi_\alpha]\,.$$

In order to prove the Jacobi identity $J(\xi,\eta,\zeta) = 0$, since J is trilinear, it is enough to consider the case $\xi \in \mathrm{gr}_\alpha G$, $\eta \in \mathrm{gr}_\beta G$ and $\zeta \in \mathrm{gr}_\gamma G$. Now using the Proposition 1.1(3) we have, for $x \in G_\alpha$, $y \in G_\beta$, $z \in G_\gamma$ such that $\bar{x} = \xi$, $\bar{y} = \eta$, $\bar{z} = \zeta$.

$$J(\xi,\eta,\zeta) = \overline{(x^y,(y,z))(y^z,(z,x))(z^x,(x,y))} = \bar{1} = 0$$

because $\overline{x^y} = \xi$, $\overline{y^z} = \eta$, $\overline{z^x} = \zeta$. q.e.d.

3. Integral filtrations of a group

Proposition 3.1. *For any group G the following two objects are in a one-one correspondence:*

1) *Filtrations* $w : G \to \mathbf{R} \cup \{+\infty\}$ *such that* $w(G) \subset \mathbf{N} \cup \{+\infty\}$.

2) *Decreasing sequences* $\{G_n\}_{n\in\mathbf{N}}$ *of subgroups of* G *such that*
(i) $G_1 = G$.
(ii) $(G_n, G_m) \subset G_{n+m}$.

Proof. (1) $\Rightarrow$ (2) is clear.

(2) $\Rightarrow$ (1). Let $x \in G$, then we define a filtration $w : G \to \mathbf{R} \cup \{+\infty\}$ by $w(x) = \sup_{x \in G_n}\{n\}$.

It is clear that $w(1) = +\infty$, $w(x) > 0$ for all $x \in G$, and $w(x) = w(x^{-1})$.

Now let $w(x) = n$, $w(y) = m$, i.e., $x \in G_n$, $y \in G_m$ and $x \notin G_{n+1}$, $y \notin G_{m+1}$. Suppose $n \leq m$, then $G_m \subset G_n$ and therefore $xy^{-1} \in G_n$, i.e.,

$$w(xy^{-1}) \geq \inf\{w(x), w(y)\} .$$

In case $n = +\infty$ or $m = +\infty$, we have obviously this inequality.

Finally the inequality $w((x,y)) \geq w(x) + w(y)$ follows from (ii). q.e.d.

Example. The descending central series of G.

Define $G_1 = G$ and by induction $G_{n+1} = (G, G_n)$. Then the sequence $\{G_n\}$ satisfies the conditions (i)–(ii) of (2) in the Proposition 3.1. Condition (i) is satisfied by definition, and we will prove (ii) by induction on n in the pair (G_n, G_m).

Let first $n = 1$, then $(G, G_m) \subset G_{m+1}$ by definition. Now suppose $n > 1$, then

$$\begin{aligned}(G_n, G_m) = ((G, G_{n-1}), G_m) &\subset (G, (G_{n-1}, G_m))(G_{n-1}, (G, G_m)) \\ &\subset (G, G_{n+m-1})(G_{n-1}, G_{m+1}) \\ &\subset G_{n+m} \cdot G_{n+m} = G_{n+m} .\end{aligned}$$

Conversely, if $\{H_n\}$ is a decreasing sequence of subgroups of G which verifies (2), then $H_n \supset G_n$ for all n. The proof of this is also by induction. Suppose $n = 1$, then by definition $H_1 = G_1$. Now if $n \geq 1$, we have

$$H_{n+1} \supset (H_1, H_n) \supset (G, G_n) = G_{n+1} .$$

4. Filtrations in GL(n)

Let k be a field with an ultrametric absolute value $|x| = a^{v(x)}$. Let A_v be the ring of v and let $\mathfrak{m}_v$ be the maximal ideal of A_v, let $k(v) = A_v/\mathfrak{m}_v$.

Let n be a positive integer and let G be the group of $n \times n$-matrices with coefficients in A_v such that $g \equiv 1 \bmod \mathfrak{m}_v$, i.e., if $g = (g_{ij})$ then $g_{ij} \equiv \delta_{ij} \bmod \mathfrak{m}_v$.

If $g \in G$ then $g = 1 + x$ where x is a matrix with coefficients in $\mathfrak{m}_v$.

Clearly G is a group, because it can be described as

$$G = \mathrm{Ker}\{\ \mathrm{GL}(n, A_v) \to \mathrm{GL}(n, k(v))\ \} .$$

Let $X \in M_n(k)$, $X = (x_{ij})$, then define $v(X) = \inf\{v(x_{ij})\}$.

We can define a map $w : G \to \mathbf{R} \cup \{+\infty\}$ by $w(g) = v(x)$, where $g = 1 + x$.

Theorem 4.1. *The map w is a filtration on G.*

Proof. The conditions $w(1) = +\infty$ and $w(g) > 0$ for all $g \in G$ are obvious.
Let now $G_\lambda = \{ g \in G \mid w(g) \geq \lambda \}$. If $\mathfrak{a}_\lambda$ is defined by

$$\mathfrak{a}_\lambda = \{ x \mid x \in k,\ v(x) \geq \lambda \} ,$$

the set G_λ is the kernel of the canonical homomorphism

$$\mathrm{GL}(n, A_v) \to \mathrm{GL}(n, A_v/\mathfrak{a}_\lambda) .$$

Hence G_λ is a subgroup of G, and this proves condition (3).

To prove condition (4), i.e., $(G_\lambda, G_\mu) \subset G_{\lambda+\mu}$, write $g \in G_\lambda$, $h \in G_\mu$ in the form:

$$g = 1 + x , \qquad h = 1 + y .$$

One must check that $hg \equiv gh \pmod{G_{\lambda+\mu}}$. But

$$hg = 1 + x + y + yx$$
$$gh = 1 + x + y + xy$$

and the coefficients of xy and yx belong to $\mathfrak{a}_{\lambda+\mu}$. Hence hg and gh have the same image in $\mathrm{GL}(n, A_v/\mathfrak{a}_{\lambda+\mu})$, and they are congruent $\bmod G_{\lambda+\mu}$, q.e.d.

Exercises

1. Determine the Lie algebra $\mathrm{gr}\, G$.
2. Prove that $G = \varprojlim G/G_\lambda$ if k is complete.

Chapter III. Universal Algebra of a Lie Algebra

1. Definition

Let k be a commutative ring and let $\mathfrak{g}$ be a Lie algebra over k.

Definition 1.1. A *universal algebra* of $\mathfrak{g}$ is a map $\varepsilon : \mathfrak{g} \to U\mathfrak{g}$, where $U\mathfrak{g}$ is an associative algebra, with a unit satisfying the following properties:

1). ε is a Lie algebra homomorphism,

$$\text{(i.e., } \varepsilon \text{ is } k\text{-linear and } \varepsilon[x,y] = \varepsilon x \cdot \varepsilon y - \varepsilon y \cdot \varepsilon x).$$

2). If A is any associative algebra with a unit and $\alpha : \mathfrak{g} \to A$ is any Lie algebra homomorphism, there is a unique homomorphism of associative algebras $\varphi : U\mathfrak{g} \to A$ such that the diagram

$$\begin{array}{ccc} \mathfrak{g} & \xrightarrow{\varepsilon} & U\mathfrak{g} \\ {\scriptstyle\alpha}\downarrow & \swarrow {\scriptstyle\varphi} & \\ A & & \end{array}$$

is commutative [i.e., there is an isomorphism

$$\mathrm{Hom}_{\mathrm{Lie}}(\mathfrak{g}, LA) \cong \mathrm{Hom}_{\mathrm{Ass}}(U\mathfrak{g}, A)$$

where LA is the Lie algebra associated to A, cf. Chap. I, example (iii).]

It is trivial that $U\mathfrak{g}$, if it exists, is unique (up to a unique isomorphism). To prove its existence, we use the *tensor algebra* $T\mathfrak{g}$ of $\mathfrak{g}$, i.e., $T\mathfrak{g} = \sum_{n=0}^{\infty} T^n\mathfrak{g}$, where $T^n\mathfrak{g} = \mathfrak{g} \otimes \cdots \otimes \mathfrak{g} = \bigotimes^n \mathfrak{g}$ for $n \geq 0$. For any associative algebra A with a unit, one has: $\mathrm{Hom}_{\mathrm{Mod}}(\mathfrak{g}, A) = \mathrm{Hom}_{\mathrm{Ass}}(T\mathfrak{g}, A)$.

Now let I be the two-sided ideal of $T\mathfrak{g}$ generated by the elements of the form $[x,y] - x \otimes y + y \otimes x, \ x, y \in \mathfrak{g}$.

Take $U\mathfrak{g} = T\mathfrak{g}/I$, then we have:

Theorem 1.2. *Let $\varepsilon : \mathfrak{g} \to U\mathfrak{g}$ be the composition $\mathfrak{g} \to T^1\mathfrak{g} \to T\mathfrak{g} \to U\mathfrak{g}$. Then the pair $(U\mathfrak{g}, \varepsilon)$ is a universal algebra of $\mathfrak{g}$.*

In fact, let α be a Lie homomorphism of $\mathfrak{g}$ into an associative algebra A. Since α is k-linear, it extends to a unique homomorphism $\psi : T\mathfrak{g} \to A$. It is clear that $\psi(I) = 0$, hence ψ defines $\varphi : U\mathfrak{g} \to A$, and we have checked the universal property of $U\mathfrak{g}$.

Remark. Let E be a $\mathfrak{g}$-module (i.e., a k-module with a bilinear product $\mathfrak{g} \times E \to E$ such that $[x,y]e = x(ye) - y(x \cdot e)$ for $x, y \in \mathfrak{g}$, $e \in E$). The map $\mathfrak{g} \to \mathrm{End}(E,E)$ which defines the module structure of E is a Lie homomorphism. Hence it extends to an algebra homomorphism $U\mathfrak{g} \to \mathrm{End}(E,E)$ and E becomes a $U\mathfrak{g}$-*left-module.* It is easy to check that one obtains in this way an *isomorphism* of the category of $\mathfrak{g}$-modules onto the category of $U\mathfrak{g}$-*left-modules.*

Exercise (Bergman). Prove that $U\mathfrak{g} = k \iff \mathfrak{g} = 0$. (Hint: use the adjoint representation.)

2. Functorial properties

1). If $\mathfrak{g} = \varinjlim \mathfrak{g}_\alpha$, then $U\mathfrak{g} = \varinjlim U\mathfrak{g}_\alpha$.

2). If $\mathfrak{g} = \mathfrak{g}_1 \times \mathfrak{g}_2$, where $\mathfrak{g}_1$ and $\mathfrak{g}_2$ commute, then $U\mathfrak{g} = U\mathfrak{g}_1 \otimes U\mathfrak{g}_2$.

3). Let k' be an extension of k and let $\mathfrak{g}' = \mathfrak{g} \otimes_k k'$, then $U\mathfrak{g}' = U\mathfrak{g} \otimes_k k'$.

Proof of 2). Consider the homomorphisms $\varepsilon_i : \mathfrak{g}_i \to U\mathfrak{g}_i$, $i = 1, 2$, $f : \mathfrak{g} \to U\mathfrak{g}_1 \otimes U\mathfrak{g}_2$ given by $f(x) = \varepsilon(x_1) \otimes 1 + 1 \otimes \varepsilon(x_2)$ where $x = x_1 + x_2$ with $x_1 \in \mathfrak{g}_1$, $x_2 \in \mathfrak{g}_2$. The map f is a Lie algebra homomorphism since $\mathfrak{g}_1$ commutes with $\mathfrak{g}_2$. Hence f induces an associative algebra homomorphism $\psi : U\mathfrak{g} \to U\mathfrak{g}_1 \otimes U\mathfrak{g}_2$.

On the other hand we have the homomorphisms $\mathfrak{g}_i \to \mathfrak{g} \to U\mathfrak{g}$, $i = 1, 2$, which induce homomorphisms $\varphi_1 : U\mathfrak{g}_i \to U\mathfrak{g}$ and since $\mathfrak{g}_1$ commutes with $\mathfrak{g}_2$ we have that $\varphi_1(x_1)\varphi_2(x_2) = \varphi_2(x_2)\varphi_1(x_1)$ for all $x_1 \in \mathfrak{g}_1$, $x_2 \in \mathfrak{g}_2$.

Finally take $\varphi : U\mathfrak{g}_1 \otimes U\mathfrak{g}_2 \to U\mathfrak{g}$ given by $\varphi(x_1 \otimes x_2) = \varphi_1(x_1)\varphi_2(x_2)$, then we have $\psi \circ \varphi = \mathrm{id}$ and $\varphi \circ \psi = \mathrm{id}$.

The proof of 1) and 3) are similar.

3. Symmetric algebra of a module

Let $\mathfrak{g}$ be a k-module and define $[x, y] = 0$ for all $x, y \in \mathfrak{g}$. In this case, the universal algebra $U\mathfrak{g}$ of $\mathfrak{g}$ is called the *symmetric algebra* of the k-module $\mathfrak{g}$ and it is denoted by $S\mathfrak{g}$.

We can define $S\mathfrak{g}$ as the largest commutative quotient of $T\mathfrak{g}$, i.e., $S\mathfrak{g} = \sum_{n=0}^{\infty} S^n\mathfrak{g}$ where $S^n\mathfrak{g} = (\bigotimes^n \mathfrak{g})/I$ where I is generated by the elements of the form $a - \sigma a$ where σ is a permutation of $[1, n]$, and $a \in \bigotimes^n \mathfrak{g}$.

We will consider the case where $\mathfrak{g}$ is a free k-module with basis $(e_i)_{i \in I}$.

Let $\varepsilon : \mathfrak{g} \to k[(X_i)_{i \in I}]$ be the homomorphism given by $\varepsilon(e_i) = X_i$ where $k[(X_i)_{i \in I}]$ is the polynomial ring in the indeterminates X_i, $i \in I$. Then $(\varepsilon, k[(X_i)_{i \in I}])$ has the universal property of 1.1, i.e., ε is a k-linear map such that $\varepsilon(x)\varepsilon(y) = \varepsilon(y)\varepsilon(x)$ and if $f : \mathfrak{g} \to A$ is a k-linear map with $f(x)f(y) = f(y)f(x)$ for all $x, y \in \mathfrak{g}$ where A is an associative algebra, then there exists an associative algebra homomorphism $f^* : k[(X_i)] \to A$ such that $f^* \circ \varepsilon = f$. In fact if $P(x_i) \in k[(X_i)]$ then $f^*(P) = P(f(e_i))$. This shows that we can identify $S\mathfrak{g}$ with the polynomial algebra $k[(X_i)_{i \in I}]$.

If we assume that I is totally ordered, then $S\mathfrak{g}$ has for basis the set of monomials $e_{i_1} \cdots e_{i_n}$, $i_1 \leq i_2 \leq \cdots \leq i_n$, $n \geq 0$.

4. Filtration of $U\mathfrak{g}$

Let $\mathfrak{g}$ be a Lie algebra over k, and let $U\mathfrak{g}$ be the universal algebra of $\mathfrak{g}$. We define a filtration of $U\mathfrak{g}$ as follows: Let $U_n\mathfrak{g}$ be the submodule of $U\mathfrak{g}$ generated by the products $\varepsilon(x_1)\cdots\varepsilon(x_m)$, $m \le n$, where $x_i \in \mathfrak{g}$. We have

$$U_0\mathfrak{g} = k$$
$$U_1\mathfrak{g} = k \oplus \varepsilon(\mathfrak{g})$$

and $U_0\mathfrak{g} \subset U_1\mathfrak{g} \subset \cdots \subset U_n\mathfrak{g} \subset U_{n+1}\mathfrak{g} \subset \cdots$.

Now we define $\operatorname{gr} U\mathfrak{g} = \sum_{n=0}^{\infty} \operatorname{gr}_n U\mathfrak{g}$, where $\operatorname{gr}_n U\mathfrak{g} = U_n\mathfrak{g}/U_{n-1}\mathfrak{g}$.

The map $U_p\mathfrak{g} \times U_q\mathfrak{g} \to U_{p+q}\mathfrak{g}$ given by $(a,b) \mapsto ab$ defines, by passage to quotient, a bilinear map

$$\operatorname{gr}_p U\mathfrak{g} \times \operatorname{gr}_q U\mathfrak{g} \to \operatorname{gr}_{p+q} U\mathfrak{g} \ .$$

We then obtain a structure of *graded algebra* on $\operatorname{gr} U\mathfrak{g}$; with this structure $\operatorname{gr} U\mathfrak{g}$ is called the *graded algebra* associated to $U\mathfrak{g}$. It is associative and has a unit.

Proposition 4.1. *The algebra* $\operatorname{gr} U\mathfrak{g}$ *is generated by the image of* $\mathfrak{g}$ *under the map induced by* $\varepsilon : \mathfrak{g} \to U\mathfrak{g}$.

Proof. Let $\alpha \in \operatorname{gr}_n U\mathfrak{g}$ and let $a \in U_n\mathfrak{g}$ be a representative of α, i.e., $\bar{a} = \alpha$. Now, we have $a = \sum_{m_\mu \le n} \lambda_\mu \varepsilon(x_1^{(\mu)}) \cdots \varepsilon(x_{m_\mu}^{(\mu)})$. Thus we have

$$\alpha = \sum_{m_\mu = n} \lambda_\mu \overline{\varepsilon(x_1^{(\mu)})} \cdots \overline{\varepsilon(x_{m_\mu}^{(\mu)})} \qquad \text{q.e.d.}$$

Theorem 4.2. *The algebra* $\operatorname{gr} U\mathfrak{g}$ *is commutative.*

Proof. Using 4.1 it is enough to prove that $\overline{\varepsilon(x)}$, $\overline{\varepsilon(y)}$ commute in $\operatorname{gr}_2 U\mathfrak{g}$ for all $x, y \in \mathfrak{g}$.

Since ε is a Lie algebra homomorphism we have

$$\varepsilon(x)\varepsilon(y) - \varepsilon(y)\varepsilon(x) = \varepsilon([x,y]) \ ,$$

but $\varepsilon([x,y]) \in U_1\mathfrak{g}$ so $\varepsilon(x)\varepsilon(y) \equiv \varepsilon(y)\varepsilon(x) \bmod U_1\mathfrak{g}$. Therefore

$$\overline{\varepsilon(x)}\,\overline{\varepsilon(y)} = \overline{\varepsilon(y)}\,\overline{\varepsilon(x)} \ .$$

It follows from Theorem 4.2 that the canonical map $\mathfrak{g} \to \operatorname{gr} U\mathfrak{g}$ extends to a homomorphism

$$\imath : S\mathfrak{g} \to \operatorname{gr} U\mathfrak{g}$$

where $S\mathfrak{g}$ is the symmetric algebra of $\mathfrak{g}$ (cf. III.3).

Since $\operatorname{gr} U\mathfrak{g}$ is generated by the image of $\mathfrak{g}$, $\imath$ *is surjective.*

Theorem 4.3 (Poincaré-Birkhoff-Witt). *If* $\mathfrak{g}$ *is a k-free module, then* $\imath$ *is an isomorphism.*

In order to prove the theorem we will prove first two lemmas.

Let $(x_i)_{i\in I}$ be a basis of $\mathfrak{g}$ and choose a total order in I.

Lemma 4.4. *The family of monomials* $\varepsilon(x_{i_1})\cdots\varepsilon(x_{i_m})$, $i_1 \leq \cdots \leq i_m$, $m \leq n$, *generate* $U^n\mathfrak{g}$ *(as a k-module).*

Proof. We proceed by induction with respect to n.

For $n = 0$ the statement is trivial.

Suppose now $n > 0$ and take $a \in U^n\mathfrak{g}$. Then its image $\bar{a} \in \mathrm{gr}^n U\mathfrak{g}$ is a polynomial of degree n in the $\varepsilon(x_i)$, but this implies a is a linear combination of products $\varepsilon(x_{i_1})\cdots\varepsilon(x_{i_n})$, $i_1 \leq \cdots \leq i_n$ plus an element $a_1 \in U^{n-1}\mathfrak{g}$.

By the hypothesis of induction a_1 is a linear combination of products $\varepsilon(x_{i_1})\cdots\varepsilon(x_{i_m})$, $i_1 \leq \cdots \leq i_m$, $m < n$. q.e.d.

Lemma 4.5. *The following statement is equivalent to 4.3:*

The family of monomials $\varepsilon(x_{i_1})\cdots\varepsilon(x_{i_n})$, $i_1 \leq \cdots \leq i_n$, $n \geq 0$ *is a basis of* $U\mathfrak{g}$.

For $M = (i_1, \ldots, i_m)$ with $i_1 \leq i_2 \leq \cdots \leq i_m$, write

$$x_M = \varepsilon(x_{i_1})\cdots\varepsilon(x_{i_m}),$$

and denote the *length* of M by $\ell(M) = m$. For each $n \geq 0$ the elements x_M with $\ell(M) = n$ lie in $U_n\mathfrak{g}$, and their images $\bar{x}_M$ in $\mathrm{gr}_n U\mathfrak{g} = U_n\mathfrak{g}/U_{n-1}\mathfrak{g}$ are the images, under the map $\imath : S^n\mathfrak{g} \to \mathrm{gr}_n U\mathfrak{g}$, of the monomial basis elements of $S^n\mathfrak{g}$. Thus, the injectivity of $\imath$ is equivalent to the non-existence of a relation

$$\sum_{\ell(M)=n} c_M x_M \equiv 0 \pmod{U_{n-1}\mathfrak{g}}$$

with some $c_M \neq 0$. By Lemma 4.4 this is the same as the non-existence of a relation

$$\sum_{\ell(M)=n} c_M x_M = \sum_{\ell(M)<n} c_M x_M ,$$

with some c_M on the left not zero. But any non-trivial k-linear dependence relation among the x_M can be put in the latter form. Hence Lemma 4.5 is true, and we can now proceed to prove Theorem 4.3 in the new form.

To do so we can (and will) assume that I is *well-ordered.* Let V be the free k-module with basis $\{z_M\}$ where M runs through the set of all sequences $(i_1, \ldots, i_n)$ with $n \geq 0$ and $i_1 \leq i_2 \cdots \leq i_n$ as above. If $i \in I$ and $M = (i_1, \ldots, i_n)$, we define $i \leq M \iff i \leq i_1$, in which case we introduce the notation $iM = (i, i_1, \ldots, i_n)$.

Main lemma. *We can make* V *into a* $\mathfrak{g}$*-module in such a way that* $x_i Z_M = Z_{iM}$ *whenever* $i \leq M$.

We shall first define a k-bilinear map $(x, v) \mapsto xv$ of $\mathfrak{g} \times V$ into V, and will then prove that it makes V a $\mathfrak{g}$-module, that is, satisfies

$$(1) \qquad xyv - yxv = [x, y]v\,, \qquad \text{for } x, y \in \mathfrak{g}, \text{ and } v \in V.$$

To define xv it suffices to define $x_i Z_M$ for all i and M, and to define $x_i Z_M$ we may assume by induction that $x_j Z_N$ has been defined for all $j \in I$ when $\ell(N) < \ell(M)$ and for $j < i$ when $\ell(N) = \ell(M)$. Moreover we assume that this has been done in such a way that the following holds:

(∗) $x_j Z_N$ is a k-linear combination of Z_L's with $\ell(L) \leq \ell(N) + 1$.

We then put

$$(2) \qquad x_i Z_M = \begin{cases} Z_{iM} & , \text{if } i \leq M \\ x_j(x_i Z_N) + [x_i, x_j] Z_N & , \text{if } M = jN \text{ with } i > j. \end{cases}$$

This makes sense because, in the second case, $x_i Z_N$ is already defined as a linear combination of Z_L's with $\ell(L) \leq \ell(N)+1 = \ell(M)$, and $[x_i, x_j]$ is a linear combination of x_k. Moreover the condition (∗) holds with j and N replaced by i and M.

To check (1) it suffices, by linearity, to show

$$(1') \qquad x_i x_j Z_N - x_j x_i Z_N = [x_i, x_j] Z_N$$

for all i, j and N. Since both sides are skew symmetric and vanish when $i = j$, we may assume $i > j$. If $j \leq N$, then $x_j Z_N = Z_{jN}$ and (1′) follows from the second case of our inductive definition (2) above. There remains the case $N = kL$, with $i > j > k$, when (1′) becomes

$$(ijk) \qquad x_i x_j x_k Z_L - x_j x_i x_k Z_L = [x_i, x_j] x_k Z_L\,.$$

By induction on $\inf(i, j)$, we know this equation does hold if we permute ijk cyclically, that is the equations (jki) and (kij) are correct. On the other hand, by induction on $\ell(N)$ we can assume $xyZ_L = yxZ_L + [x, y]Z_L$ for all $x, y \in \mathfrak{g}$. Thus the right hand side of (ijk) can be rewritten:

$$\begin{aligned} [x_i, x_j] x_k Z_L &= x_k [x_i, x_j] Z_L + [[x_i, x_j], x_k] Z_L \\ &= x_k x_i x_j Z_L - x_k x_j x_i Z_L + [[x_i, x_j], x_k] Z_L\,. \end{aligned}$$

If we substitute this on the right side of (ijk) and then add the three equations $(ijk) + (jki) + (kij)$ we get an equation of the form

$$\sum = \sum + \mathrm{Jac}(x_i, x_j, x_k) Z_L\,.$$

Hence, (ijk) is true, and our main lemma is proved.

Since V is a $\mathfrak{g}$-module, it is also a $U\mathfrak{g}$-left module, cf. Remark at the end of III.1.

In particular we have in V the element $Z_\emptyset$ where $\emptyset$ is the empty set. For all M we have $x_M Z_\emptyset = Z_M$. We will prove this by induction on $\ell(M)$. If

$\ell(M) = 0$ then it is clear because $x_M = 1$. If $\ell(M) > 0$ we write $M = iN$, $i \leq N$. Then $x_M = x_i x_N$ and $x_M Z_\emptyset = x_i x_N Z_\emptyset = x_i Z_N = Z_{iN} = Z_M$.

Finally, suppose we have $\sum c_M x_M = 0$, then

$$0 = \sum c_M x_M Z_\emptyset = \sum c_M Z_M \,,$$

but this implies $c_M = 0$ for all M. q.e.d.

Corollary 1. *If $\mathfrak{g}$ is a free k-module then $\varepsilon : \mathfrak{g} \to U\mathfrak{g}$ is injective.*

In fact, in this case $\mathfrak{g} \cong \mathrm{gr}_1\, U\mathfrak{g}$.

Corollary 2. *Let $\mathfrak{g} = \mathfrak{g}_1 \oplus \mathfrak{g}_2$ where $\mathfrak{g}_1$ and $\mathfrak{g}_2$ are subalgebras of $\mathfrak{g}$ and are free k-modules. Then the map $U\mathfrak{g}_1 \otimes U\mathfrak{g}_2 \to U\mathfrak{g}$ given by $u_1 \otimes u_2 \mapsto u_1 u_2$ is a k-linear isomorphism.*

Proof. Let $(x_i)_{i \in I}$, $(y_j)_{j \in J}$ be a basis of $\mathfrak{g}_1$ and $\mathfrak{g}_2$ respectively, then $\{(x_i), (x_j)\}$ is a basis of $\mathfrak{g}$. Take a total order in $I \cup J$ such that every element of I is less than every element of J. Applying 4.5 we have that the families of monomials $\{\varepsilon(x_{i_1}) \cdots \varepsilon(x_{i_n})\}$, $\{\varepsilon(y_{j_1}) \cdots \varepsilon(y_{j_m})\}$ and $\{\varepsilon(x_{i_1}) \cdots \varepsilon(x_{i_n})\varepsilon(y_{j_1}) \cdots \varepsilon(y_{j_m})\}$ for $i_1 \leq \cdots \leq i_n$ and $j_1 \leq \cdots \leq j_m$ are basis of $U\mathfrak{g}_1$, $U\mathfrak{g}_2$ and $U\mathfrak{g}$ respectively. Thus the map $U\mathfrak{g}_1 \otimes U\mathfrak{g}_2 \to U\mathfrak{g}$ given by $u_1 \otimes u_2 \mapsto u_1 u_2$ is a bijection on the basis of $U\mathfrak{g}_1 \otimes U\mathfrak{g}_2$ and $U\mathfrak{g}$. q.e.d.

Notice that in this case we have also induced an isomorphism

$$\mathrm{gr}\, U\mathfrak{g}_1 \otimes \mathrm{gr}\, U\mathfrak{g}_2 \xrightarrow{\approx} \mathrm{gr}\, U\mathfrak{g}$$

because $\mathrm{gr}\, U\mathfrak{g}_i = S\mathfrak{g}_i$ and $\mathrm{gr}\, U\mathfrak{g} = S\mathfrak{g} \simeq S\mathfrak{g}_1 \otimes S\mathfrak{g}_2$.

5. Diagonal map

Let $\mathfrak{g}$ be a Lie algebra over k and suppose $\mathfrak{g}$ is free as a k-module.

Definition 5.1. The Lie algebra homomorphism $\Delta : \mathfrak{g} \to \mathfrak{g} \times \mathfrak{g}$ given by $x \mapsto (x, x)$ induces a homomorphism of associative algebras

$$\Delta : U\mathfrak{g} \to U\mathfrak{g} \otimes U\mathfrak{g} \,,$$

which is called the *diagonal map*.

Proposition 5.2. *The diagonal map Δ is characterized by the following two conditions:*

1) *Δ is an algebra homomorphism.*

2) *$\Delta x = x \otimes 1 + 1 \otimes x$ for all $x \in \mathfrak{g}$.*

Notice that we identify $x \in \mathfrak{g}$ with its image in $U\mathfrak{g}$.

Definition 5.3. An element $\alpha \in U\mathfrak{g}$ is called *primitive* if $\Delta\alpha = \alpha \otimes 1 + 1 \otimes \alpha$.

Hence every element $x \in \mathfrak{g}$ is primitive.

Theorem 5.4. *Assume k is torsion free (as a $\mathbf{Z}$-module) and $\mathfrak{g}$ is a free k-module. Then the set of primitive elements of $U\mathfrak{g}$ coincides with $\mathfrak{g}$.*

Case 1. $\mathfrak{g}$ abelian. In this case $U\mathfrak{g}$ can be identified with the ring of polynomials $k[(X_i)]$ in variables X_i corresponding to the basis elements x_i of $\mathfrak{g}$. The diagonal map can be interpreted as a homomorphism $\Delta : k[X_i] \to k[(X_i'), (X_i'')]$ where $X_i' \sim X_i \otimes 1$ and $X_i'' \sim 1 \otimes X_i$, and is then given by $\Delta f(X_i', X_i'') = f(X_i' + X_i'')$, because it sends X_i to $X_i' + X_i''$ for each i. Thus the primitive elements $f(x) \in k[(X_i)]$ are those which satisfy $f(X_i' + X_i'') = f(X_i') + f(X_i'')$. If f is additive in this sense, then so is each homogeneous component f_n. If f is homogeneous of degree n and additive then

$$2^n f(X_i) = f(2X_i) = f(X_i + X_i) = 2f(X_i),$$

so $(2^n - 2)f = 0$. Since k is $\mathbf{Z}$-torsion free, we must have $f = 0$ if $n \neq 1$. Thus the only additive polynomials are the linear homogeneous ones.

Case 2. The general case. The map $\Delta : U\mathfrak{g} \to U\mathfrak{g} \otimes U\mathfrak{g}$ induces a map

$$\operatorname{gr}\Delta : \operatorname{gr} U\mathfrak{g} \to \operatorname{gr}(U\mathfrak{g} \otimes U\mathfrak{g}) \simeq \operatorname{gr} U(\mathfrak{g} \oplus \mathfrak{g}) \simeq \operatorname{gr} U\mathfrak{g} \otimes \operatorname{gr} U\mathfrak{g}$$

(see end of III.4). On the other hand, we have $\operatorname{gr} U\mathfrak{g} \simeq S\mathfrak{g}$, and the corresponding map $S\mathfrak{g} \to S\mathfrak{g} \otimes S\mathfrak{g}$ is the same as the one discussed in the first case, as one sees by looking at its effect on elements of the form $\bar{x} \in \operatorname{gr}_1 U\mathfrak{g}$ coming from elements $x \in \mathfrak{g}$.

Let $x \in U_n\mathfrak{g}$, and let $\bar{x}$ denote its image in $\operatorname{gr}_n U\mathfrak{g}$. If x is primitive, then $\bar{x}$ is primitive for $\operatorname{gr}\Delta$, hence, if $n > 1$, we have $\bar{x} = 0$ by case 1. Iterating this, we conclude $x \in U_1\mathfrak{g}$, that is, $x = \lambda + y$, with $\lambda \in k$, $y \in \mathfrak{g}$. Then

$$\Delta x = \lambda + y \otimes 1 + 1 \otimes y$$
$$x \otimes 1 + 1 \otimes x = \lambda + y \otimes 1 + \lambda + 1 \otimes y\ .$$

Thus, if x is primitive, then $2\lambda = \lambda$, hence $\lambda = 0$, and $x \in \mathfrak{g}$.

Exercises

1. Let $PU\mathfrak{g}$ denote the set of primitive elements of $U\mathfrak{g}$. Show that $PU\mathfrak{g}$ is stable under $[\ ,\ ]$, that is, if x and $y \in PU\mathfrak{g}$, so is $xy - yx$.
2. Suppose $pk = 0$ for some prime number p, and suppose $\mathfrak{g}$ is free, with basis $(x_i)_{i \in I}$. Show
   a) $PU\mathfrak{g}$ is stable under the map $y \mapsto y^p$.
   b) The elements $(x_i^{p^\nu})$, $i \in I$, $\nu \geq 1$, form a k-basis for $PU\mathfrak{g}$.
   c) If x and y are in $\mathfrak{g}$, then $(x + y)^p - x^p - y^p \in \mathfrak{g}$.

Chapter IV. Free Lie Algebras

In this chapter, k denotes a commutative and associative ring, with a unit. All modules and algebras are taken over k.

1. Free magmas

Definition 1.1. A set M with a map

$$M \times M \to M$$

denoted by $(x, y) \mapsto xy$ is called a *magma.*

Let X be a set and define inductively a family of sets X_n $(n \geq 1)$ as follows:

1) $X_1 = X$

2) $X_n = \coprod_{p+q=n} X_p \times X_q$ $(n \geq 2)$ (= disjoint union).

Put $M_X = \coprod_{n=1}^{\infty} X_n$ and define $M_X \times M_X \to M_X$ by means of

$$X_p \times X_q \to X_{p+q} \subset M_X ,$$

where the arrow is the canonical inclusion resulting from 2).

The magma M_X is called the *free magma on* X. An element w of M_X is called a non-associative word on X. Its length, $\ell(w)$, is the unique n such that $w \in X_n$.

Theorem 1.2. *Let N be any magma, and let $f : X \to N$ be any map. Then there exists a unique magma homomorphism $F : M_X \to N$ which extends f.*

Proof. Define F inductively by $F(u, v) = F(u) \cdot F(v)$ if $u, v \in X_p \times X_q$.

Properties of the free magma M_X:

1) M_X is generated by X.

2) $m \in M_X - X \iff m = u.v$, with $u, v \in M$; and u, v are uniquely determined by m.

2. Free algebra on X

Let A_X be the k-algebra of the free magma M_X. An element $\alpha \in A_X$ is a finite sum $\alpha = \sum_{m \in M_X} c_m m$, with $c_m \in k$; the multiplication in A_X extends the multiplication in M_X.

Definition 2.1. The algebra A_X is called the *free algebra* on X.

This definition is justified by the following:

Theorem 2.2. *Let B be a k-algebra and let $f : X \to B$ be a map. There exists a unique k-algebra homomorphism $F : A_X \to B$ which extends f.*

Proof. By 1.2, we can extend f to a magma homomorphism $f' : M_X \to B$, where B is viewed as a magma under multiplication. This map extends by linearity to a k-linear map $F : A_X \to B$. One checks easily that F is an algebra homomorphism. The uniqueness of F follows from the fact that X generates A_X.

Remark. A_X is a graded algebra, the homogeneous elements of degree n being those which are linear combinations of words $m \in M_X$ of length n.

3. Free Lie algebra on X

Let I be the two-sided ideal of A_X generated by the elements of the form aa, $a \in A_X$ and $J(a,b,c)$, where $a,b,c \in A_X$ ($J(a,b,c) = (ab)c + (bc)a + (ca)b$).

Definition 3.1. The quotient algebra A_X/I is called the *free Lie algebra* on X.

This algebra will be denoted by $L_X(k)$, or simply L_X.

Functorial properties.

1) If $f : X \to X'$ is any map, then there exists a unique map $F : L_X \to L_{X'}$ such that $F|_X = f$.

1′) If $\{X_\alpha, i_\alpha^\beta\}$ is a direct system and $X = \varinjlim X_\alpha$ then

$$\varinjlim L_{X_\alpha} = L_X \ .$$

2) Let k' be an extension of k, then

$$L_X(k') = L_X(k) \otimes_k k' \ .$$

3) I is a graded ideal of A_X, which implies L_X has a natural structure of graded algebra.

Proof. Let $I^\#$ be the set of $a \in A_X$ such that every homogeneous component of a belongs to I. Then $I^\#$ is a two-sided ideal and $I^\# \subset I$.

Now let $x \in A_X$, $x = \sum_{n=1}^{\infty} x_n$, x_n homogeneous. Then

$$x \cdot x = \sum x_n^2 + \sum_{n<m} (x_n x_m + x_m x_n) \ ,$$

but $x_n^2 \in I$, $x_n x_m + x_m x_n = (x_n + x_m)^2 - x_n^2 - x_m^2 \in I$, so that $x \cdot x \in I^\#$. For three elements, $x = \sum x_n$, $y = \sum y_n$, and $z = \sum z_n$ we have $J(x,y,z) = \sum_{l,m,n} J(x_l, y_m, z_n) \in I^\#$. Thus $I^\# = I$, q.e.d.

4) The homogeneous component L_X^1 has basis X and the homogeneous component L_X^2 has for basis the family of elements $[x,y]$, $x < y$, $x,y \in X$, where we have chosen a total order on X.

Proof. Clearly X generates L_X and $[X,X]$ generate L_X^2 ($[X,X] = \{ [x,y], x < y, x,y \in X \}$). Consider the module $E = k^{(X)}$ and the Lie algebra $E \oplus \bigwedge^2 E = \mathfrak{g}$ (example ii′ of Chapter I). The canonical map $X \to \mathfrak{g}$ induces a Lie algebra homomorphism $L_X \to \mathfrak{g}$, and the composition $L_X^1 \oplus L_X^2 \to L_X \to \mathfrak{g}$ is an isomorphism q.e.d.

4. Relation with the free associative algebra on X

Definition 4.1. Let $E = k^{(X)}$ be the free k-module with basis X. Then the free associative algebra on X, denoted by Ass_X, is the tensor algebra TE of E.

(Elements of Ass_X may be called "associative but non-commutative" polynomials in the elements of X.)

Theorem 4.2. *Let $\phi : L_X \to \mathrm{Ass}_X$ and $\Phi : UL_X \to \mathrm{Ass}_X$ be the maps induced by the map $X \to \mathrm{Ass}_X$. Then:*

1) *The map Φ is an isomorphism.*

2) *The map ϕ is an isomorphism of L_X onto the Lie subalgebra of Ass_X generated by X.*

3) *L_X and its homogeneous components L_X^n are free k-modules.*

4) *If X is finite and $\mathrm{Card}\, X = d$ then L_X^n is free of finite rank $\ell_d(n)$ and*

$$(*) \qquad \sum_{m|n} m\ell_d(m) = d^n$$

Remark. The formula $(*)$ determines $\ell_d(n)$ by induction on n. In fact,

$$n\ell_d(n) = d^n - \sum_{\substack{m|n \\ m<n}} m\ell_d(m) .$$

(More precisely, let μ be the *Möbius function*, defined by:

$$\sum_{n=1}^{\infty} \mu(n)n^{-s} = 1/\zeta(s) = \prod_p (1 - p^{-s}) .$$

One has:

$$n\ell_d(n) = \sum_{m|n} \mu(m) d^{n/m} . \;)$$

Proof of Theorem 4.2.

1) is clear: the map $X \to UL_X$ defines a homomorphism Ψ of Ass_X into UL_X, and $\Phi \circ \Psi = 1$, $\Psi \circ \Phi = 1$.

Note also that ϕ maps L_X *onto* the Lie subalgebra of Ass_X generated by X, so that (2) is equivalent to saying that ϕ is *injective*. Note also that (3) $\Rightarrow$ (2); for, if L_X is free over k, the Birkhoff-Witt theorem shows that $L_X \to UL_X$ is injective, and we can identify UL_X with Ass_X.

The rest of the proof is divided into four steps:

First step: Assume k is a field and X is finite.

Choose a homogeneous basis $(\gamma_i)_{i\in I}$ of L_X and a total order of I.

Put $d_i = \deg(\gamma_i)$.

Now the Birkhoff-Witt theorem implies that the family of elements

$$\gamma^e = \gamma_{i_1}^{e_{i_1}} \cdots \gamma_{i_s}^{e_{i_s}} \qquad \text{with } i_1 < \cdots < i_s$$

is a basis of $UL_X = \mathrm{Ass}_X$ and we have $\deg(\gamma^e) = \sum e_{i_j} d_{i_j}$.

Let $a(n)$ be the rank of Ass_X^n, then $a(n)$ is equal to the number of families (e_i) such that $n = \sum e_i d_i$.

This last statement is equivalent to the fact that the formal power series $A(t) = \sum a(n)t^n$ may be expressed in the form

$$A(t) = \prod_{i\in I} \frac{1}{1-t^{d_i}}$$

because $\prod_{i\in I} \frac{1}{1-t^{d_i}} = \prod_{i\in I}(1 + t^{d_i} + t^{2d_i} + \cdots)$ and the coefficient of t^n in this product is precisely the number of families (e_i) such that $\sum e_i d_i = n$.

Now, for any positive integer m we have that in the product $\prod_{i\in I} \frac{1}{1-t^{d_i}}$ the number of factors such that $d_i = m$ is the rank $\ell_d(m)$ of L_X^m, i.e.,

$$A(t) = \prod_{m=1}^{\infty} \frac{1}{(1-t^m)^{\ell_d(m)}} .$$

On the other hand, since Ass_X is the free associative algebra on X the family of monomials $x_{i_1} \cdots x_{i_n}$, $x_{i_\nu} \in X$ is a basis of Ass_X^n.

This implies that $a(n) = d^n$ and therefore

$$A(t) = \sum d^n t^n = \frac{1}{1-dt}$$

i.e.,

$$\prod_{m=1}^{\infty} \frac{1}{(1-t^m)^{\ell_d(m)}} = \frac{1}{1-dt} .$$

From the equality $\log \frac{1}{1-t} = \sum_{n=1}^{\infty} \frac{1}{n} t^n$ we conclude that

$$\sum_{m,\nu} \frac{1}{\nu} \ell_d(m) t^{m\nu} = \sum_{n=1}^{\infty} \frac{1}{n} d^n t^n$$

and hence, for each n, we have $\frac{1}{n}d^n = \sum_{m\nu=n} \frac{1}{\nu}\ell_d(m)$, i.e.,

$$d^n = \sum_{m|n} m\ell_d(m)$$

which proves (4) in this case.

Second Step: Assume $k = \mathbf{Z}$ and X is a finite set.

We will use the following lemma.

Lemma 4.3. *If E is a finitely generated $\mathbf{Z}$-module and $\dim(E \otimes_{\mathbf{Z}} \mathbf{F}_p)$ over $\mathbf{F}_p = \mathbf{Z}/p\mathbf{Z}$ is independent of p, for all primes p, then E is a $\mathbf{Z}$-free module with rank equal to the dimension of $E \otimes_{\mathbf{Z}} \mathbf{F}_p$ over $\mathbf{F}_p$.*

This lemma is an easy consequence of the structure theorem of abelian groups.

Now, since $L^n_X(\mathbf{Z}) \otimes_{\mathbf{Z}} \mathbf{F}_p = L^n_X(\mathbf{F}_p)$ and $\dim(L^n_X(\mathbf{F}_p)) = \ell_d(n)$ which is independent of p, it follows that L^n_X is $\mathbf{Z}$-free with rank $\ell_d(n)$.

This proves the theorem in this case.

Third Step: Assume $k = \mathbf{Z}$ and X is an arbitrary set.

Let $\{Y_\alpha\}$ be the family of finite subsets of X, then $X = \varinjlim_\alpha Y_\alpha$.

We first prove (2).

Using the second case, we have that the map

$$\phi_\alpha : L_{Y_\alpha} \to \mathrm{Ass}_{Y_\alpha}$$

is injective for all α.

Now $\phi = \varinjlim_\alpha \phi_\alpha$ and the inductive limit of a family of injective maps is injective. This proves (2).

In particular (2) implies that L_X and L^n_X are $\mathbf{Z}$-submodules of Ass_X, which is free, so L_X and L^n_X are free for all n.

This proves the theorem in the third case.

Fourth Step: General case.

The equality $L^n_X(k) = L^n_X(\mathbf{Z}) \otimes_{\mathbf{Z}} k$ together with the third case imply $L^n_X(k)$ is k-free, i.e., (3) and therefore (2) holds.

On the other hand $\mathrm{rk}\, L^n_X(k) = \mathrm{rk}\, L^n_X(\mathbf{Z})$ thus, if X is finite, (4) holds. q.e.d.

5. P. Hall families

Definition 5.1. Let X be a set. A *P. Hall family* in M_X, the free magma on X, is a totally ordered subset H of M_X such that:

(1) $X \subset H$.

(2) If $u, v \in H$ with $\ell(u) < \ell(v)$ then $u < v$.

(3) Let $u \in M_X - X$ and let $u = vw$ be the unique decomposition of u where $v, w \in M_X$. Then $u \in H$ if and only if the following two conditions are satisfied:

(a) $v \in H$, $w \in H$ and $v < w$,

(b) either $w \in X$ or $w = w'w''$ with $w' \in H$, $w'' \in H$ and $w' \leq v$.

Lemma 5.2. *There exists a P. Hall family for any set X.*

Proof. We define by induction $H^n = H \cap X_n$. We take $H^1 = X$, and choose a total order on X. Suppose now $H^1, \ldots, H^{n-1}$ have been defined and totally ordered in such a way that (1), (2), (3) hold for elements of length $\leq n-1$. The set H^n is then defined without ambiguity by condition (3); we choose any total order on H^n, and put $u < v$ if $u \in H^i$ $(i \leq n-1)$ and $v \in H^n$. This completes the induction process, and it is clear that $H = \bigcup H^n$ is a P. Hall family.

Example. Let $X = \{x, y\}$, with $x \neq y$. We can take $H^1, \ldots, H^5$ as follows:

$$\begin{aligned}
H^1 &= \{x, y\}, \qquad x < y \\
H^2 &= \{x \cdot y\} \\
H^3 &= \{x \cdot (x \cdot y), y \cdot (x \cdot y)\}, \qquad x \cdot (x \cdot y) < y \cdot (x \cdot y) \\
H^4 &= \{x(x(xy)), y(x(xy)), y(y(xy))\} \\
H^5 &= \{x(x(x(xy))), y(x(x(xy))), y(y(x(xy))), y(y(y(xy))), \\
&\qquad (xy)(x(xy)), (xy)(y(xy))\}
\end{aligned}$$

Theorem 5.3. *If H is a P. Hall family in M_X, then the canonical images of the elements $h \in H$ in L_X make up a basis of L_X.*

Let $h \in H$ and denote by $\bar{h}$ its image in L_X. Theorem 5.3 is equivalent to:

(1) The family $\{\bar{h}\}$, $h \in H$, generates L_X.

(2) The elements $\{\bar{h}\}$, $h \in H$, are linearly independent.

We prove here only the (easier) part (1). For a proof of (2), the reader may look in M. Hall, *The Theory of Groups*, p. 170–171, or E. Witt, *Die Unterringe der freien Lieschen Ringe*, Math. Zeit., 1956; M. Hall's proof is based on a counting argument; Witt's proof is better (but longer). (See also Bourbaki, LIE II, §2, nº 11.)

Proof of (1). Let L'_X be the k-module generated by $\bar{h}$; since L'_X contains X, it will be enough to show that L'_X is a *Lie algebra*, i.e., that $h_1, h_2 \in H$ implies that $[\bar{h}_1, \bar{h}_2]$ is in L'_X.

We will carry the proof by a double induction, first on the length of h_1 + length of h_2 (which is the length n of $h_1 h_2$) and finally for a given n, by decreasing induction on $\mathrm{Inf}(h_1, h_2)$; in order that this induction process work, we will assume that X is *finite*; the general case will follow by passing to an inductive limit.

We may suppose $h_1 < h_2$ (otherwise we use the relations $[\bar{h}_1, \bar{h}_2] = -[\bar{h}_2, \bar{h}_1]$ and $[\bar{h}, \bar{h}] = 0$).

First Case. Let $h_2 \in X$, then $h_1 \in X$ since $h_1 < h_2$, so we have $h_1 h_2 \in H$ and therefore $\overline{h_1 h_2} = [\bar{h}_1, \bar{h}_2]$, q.e.d.

Second Case. $h_2 \notin X$. Put $h_2 = h_3 h_4$, $h_3, h_4 \in H$ and $h_3 < h_4$.

We have the following subcases:

a) $h_3 \leq h_1$ and then $h_1(h_3 h_4) \in H$, so

$$[\bar{h}_1, \bar{h}_2] = [\bar{h}_1, [\bar{h}_3, \bar{h}_4]] = \overline{h_1(h_3 h_4)} \, .$$

b) $h_1 < h_3 < h_4$. Using the Jacobi identity we get

$$[\bar{h}_1, [\bar{h}_3, \bar{h}_4]] = [\bar{h}_3, [\bar{h}_1, \bar{h}_4]] - [\bar{h}_4, [\bar{h}_1, \bar{h}_3]] \, .$$

Now length of $h_1 h_4 <$ length of $h_1 h_2$, hence we can apply the induction hypothesis, i.e., $[\bar{h}_1, \bar{h}_4] = \sum c_\alpha \bar{h}_\alpha$ where $h_\alpha \in H$.

From this equality we get $\ell(h_\alpha) = \ell(h_1) + \ell(h_4)$ which implies $\ell(h_\alpha) > \ell(h_1)$, hence $h_\alpha > h_1$. Since we have $h_1 < h_3$, we obtain $\mathrm{Inf}(h_3, h_\alpha) > h_1 = \mathrm{Inf}(h_1, h_2)$.

Applying the induction hypothesis we see that $[\bar{h}_3, \bar{h}_\alpha]$ is a linear combination of $\bar{h}$'s with $h \in H$.

Similarly, replacing h_3 by h_4, we see that $[\bar{h}_4, [\bar{h}_1, \bar{h}_3]]$ is also a linear combination of $\bar{h}$'s with $h \in H$. q.e.d.

6. Free groups

(In this section, we take $k = \mathbf{Z}$.)

Let X be a set and let F_X be the free group on X. Let F_X^n be the descending central series of F_X, defined by $F_X^1 = F_X$ and $F_X^n = (F_X, F_X^{n-1})$, for $n > 1$.

The associated graded group is, as we know, a Lie algebra, given by

$$\mathrm{gr}\, F_X = \sum_{n=1}^{\infty} \mathrm{gr}^n\, F_X \, , \qquad \mathrm{gr}^n\, F_X = F_X^n / F_X^{n+1} \, .$$

In particular, $\mathrm{gr}^1\, F_X = F_X/(F_X, F_X)$, that is, $\mathrm{gr}^1\, F_X$ is the free abelian group on X.

Theorem 6.1. *The canonical map $X \to \mathrm{gr}^1\, F_X$ induces an isomorphism of Lie algebras*

$$\phi_1 : L_X \xrightarrow{\sim} \mathrm{gr}\, F_X \, .$$

Corollary 6.2. *The groups F_X^n / F_X^{n+1} are free $\mathbf{Z}$-modules and if X is finite with* $\mathrm{Card}\, X = d$, *then* $\mathrm{rk}(F_X^n / F_X^{n+1}) = \ell_d(n)$.

Now consider the free associative algebra Ass_X on X; let Ass_X^n the component of degree n of Ass_X. The *completion* $\widehat{\mathrm{Ass}}_X$ of Ass_X is defined as the infinite product $\prod_{n=0}^{\infty} \mathrm{Ass}_X^n$. An element $f \in \widehat{\mathrm{Ass}}_X$ can be represented by a formal series $f = \sum_{n=0}^{\infty} f_n$, with $f_n \in \mathrm{Ass}_X^n$.

Define a homomorphism $\theta : F_X \to \widehat{\mathrm{Ass}}_X^*$ by $\theta(x) = 1 + x$ where $\widehat{\mathrm{Ass}}_X^*$ is the multiplicative group of the invertible elements of $\widehat{\mathrm{Ass}}_X$ (it is clear that $1 + x$ is invertible in $\widehat{\mathrm{Ass}}_X$, so it is in the multiplicative group $\widehat{\mathrm{Ass}}_X^*$).

For any positive integer n, define $\hat{\mathfrak{m}}^n \subset \widehat{\mathrm{Ass}}_X$ as

$$\hat{\mathfrak{m}}^n = \{ f \mid f = \sum_{n=0}^{\infty} f_m \text{ and } f_0 = f_1 = \cdots = f_{n-1} = 0 \} ,$$

and put $'F_X^n = \theta^{-1}(1 + \hat{\mathfrak{m}}^n)$. Then $g \in F_X$ is in $'F_X^n$ if and only if $\theta(g) = 1 + \sum_{m \geq n} \psi_n$.

Notice that $'F_X^1 = F_X$ and $'F_X^n \subset {}'F_X^{n-1}$.

Theorem 6.3. $'F_X^n = F_X^n$.

We now prove Theorems 6.1 and 6.3.

a) It is clear that $\phi_1 : L_X \to \mathrm{gr}\, F_X$ is *surjective.*

b) $('F_X^n)$ *is a filtration of* F_X. In fact, we only have to check

$$('F_X^m, {}'F_X^p) \subset {}'F_X^{m+p} .$$

To prove this, take $g \in {}'F_X^m$, $h \in {}'F_X^p$ with $\theta(g) = 1 + G$, $G \in \hat{\mathfrak{m}}^m$, $\theta(h) = 1 + H$, $H \in \hat{\mathfrak{m}}^p$.

We have $gh = hg(g, h)$ and

$$\theta(gh) = 1 + G + H + GH$$
$$\theta(hg) = 1 + G + H + HG .$$

Since θ is a homomorphism we get $\theta(gh) = \theta(hg)\theta((g, h))$, i.e.,

$$(*) \qquad \theta((g, h)) = 1 + (GH - HG) + \text{higher terms.}$$

Therefore $(g, h) \in {}'F_X^{m+p}$.

There is a natural map $\eta : {}'\mathrm{gr}\, F_X \to \mathrm{Ass}_X$ defined as follows: let $\xi \in {}'\mathrm{gr}^n\, F_X$, let $g \in {}'F_X^n$ be a representative of ξ, and let

$$\theta(g) = 1 + G_n + G_{n+1} + \cdots , \qquad \text{with } G_p \in \mathrm{Ass}_X^p.$$

We define $\eta(\xi)$ by:

$$\eta(\xi) = G_n .$$

It is easy to see that this definition does not depend on the choice of the representative g. Formula $(*)$ shows that $\eta : {}'\mathrm{gr}\, F_X \to \mathrm{Ass}_X$ is a *Lie algebra homomorphism.*

Since $'F_X^n$ is a filtration we know that $F_X^n \subset {'F_X^n}$, which induces a homomorphism $\psi : \operatorname{gr} F_X \to {'\operatorname{gr}} F_X$.

Now let us look at the composition

$$L_X \xrightarrow{\phi_1} \operatorname{gr} F_X \xrightarrow{\psi} {'\operatorname{gr}} F_X \xrightarrow{\eta} \mathrm{Ass}_X$$

where ϕ_1 is surjective and η is injective.

This composition is obviously the map $\phi : L_X \to \mathrm{Ass}_X$ given in the Theorem 4.2 and we know it is injective.

Hence ϕ_1 is injective and therefore is an isomorphism; which proves Theorem 6.1.

This implies now that ψ is injective. Let us prove, by induction, that $F_X^n = {'F_X^n}$.

If $n = 1$ then $F_X^1 = {'F_X}$ by definition.

Now suppose $n > 1$, then we have

$$F_X^n \subset {'F_X^n} \subset F_X^{n-1} = {'F_X^{n-1}}$$

and the injection $\operatorname{gr}^{n-1} F_X \to {'\operatorname{gr}^{n-1}} F_X$ is the canonical map

$$F_X^{n-1}/F_X^n \to F_X^{n-1}/{'F_X^n} ,$$

which implies $F_X^n = {'F_X^n}$. q.e.d.

7. The Campbell-Hausdorff formula

In IV.7 and IV.8, the ground ring k is supposed to be a $\mathbf{Q}$-algebra (for instance, a field of characteristic zero).

Theorem 7.1. *Let X be a set; then the free Lie algebra L_X on X coincides with the set of primitive elements of* Ass_X
(i.e., $L_X = \{ w \in \mathrm{Ass}_X \mid \Delta w = w \otimes 1 + 1 \otimes w \}$, where $\Delta : \mathrm{Ass}_X \to \mathrm{Ass}_X \otimes \mathrm{Ass}_X$ is the diagonal map).

This follows from a theorem proved in Chapter III, since Ass_X may be identified with UL_X.

Define now, as in IV.6, the completion $\widehat{\mathrm{Ass}}_X$ of Ass_X and the completion $\hat{L}_X$ of L_X by:

$$\widehat{\mathrm{Ass}_X} = \prod_{n=0}^{\infty} \mathrm{Ass}_X^n , \qquad \hat{L}_X = \prod_{n=0}^{\infty} L_X^n .$$

Define similarly the *completed tensor product* $\widehat{\mathrm{Ass}_X} \hat{\otimes} \widehat{\mathrm{Ass}_X}$ by:

$$\widehat{\mathrm{Ass}} \hat{\otimes} \widehat{\mathrm{Ass}_X} = \prod_{p,q} \mathrm{Ass}_X^p \otimes \mathrm{Ass}_X^q .$$

The diagonal map Δ extends to a map $\Delta : \widehat{\mathrm{Ass}_X} \to \widehat{\mathrm{Ass}_X} \hat{\otimes} \widehat{\mathrm{Ass}_X}$ and it is clear that Theorem 7.1 remains valid when Ass_X and $\mathrm{Ass}_X \otimes \mathrm{Ass}_X$ are replaced by their completions.

Theorem 7.2. *Let* $\hat{\mathfrak{m}} \subset \widehat{\mathrm{Ass}_X}$ *be the ideal generated by* X. *Define maps*

$$\exp : \hat{\mathfrak{m}} \to 1 + \hat{\mathfrak{m}} \qquad \textit{and} \qquad \log : 1 + \hat{\mathfrak{m}} \to \hat{\mathfrak{m}}$$

by the usual formulae:

$$\exp(x) = \sum x^n/n! \,, \qquad \log(1+x) = \sum_{n=1}^{\infty} (-1)^{n+1} x^n/n \,.$$

Then $\exp \circ \log = \mathrm{id}$ *and* $\log \circ \exp = \mathrm{id}$.

Proof. Let us prove, for instance, that $\exp(\log(1+y)) = 1 + y$ if $y \in \hat{\mathfrak{m}}$. If T is an indeterminate, the formula $\exp(\log(1+T)) = 1 + T$ is known to be true in the power series ring $\mathbf{Q}[[T]]$. But, since y belongs to $\hat{\mathfrak{m}}$, there is a well-defined and continuous homomorphism $f : \mathbf{Q}[[T]] \to \widehat{\mathrm{Ass}_X}$ which transforms T into y. Applying f to the equality $\exp(\log(1+T)) = 1 + T$, we get $\exp(\log(1+y)) = 1 + y$, q.e.d.

Corollary 7.3. *The map* exp *defines a bijection of the set of* $\alpha \in \hat{\mathfrak{m}}$ *with* $\Delta\alpha = \alpha \otimes 1 + 1 \otimes \alpha$ *onto the set of* $\beta \in 1 + \hat{\mathfrak{m}}$ *with* $\Delta\beta = \beta \otimes \beta$.

Proof. Let $\alpha \in \hat{\mathfrak{m}}$ and $\beta = e^\alpha$. Since Δ commutes with the exponential map and $\alpha \otimes 1$ commutes with $1 \otimes \alpha$, we obtain

$$\begin{aligned} \Delta\beta &= \Delta e^\alpha = e^{\Delta\alpha} = e^{\alpha\otimes 1 + 1\otimes\alpha} = e^{\alpha\otimes 1} e^{1\otimes\alpha} = (\beta\otimes 1)(1\otimes\beta) \\ &= \beta \otimes \beta \,. \end{aligned}$$

Theorem 7.4 (Campbell-Hausdorff). *Let* $X = \{x, y\}$, $x \neq y$, *then* $e^x e^y = e^z$ *with* $z \in \hat{L}_X$.

Proof. Since $e^x, e^y \in 1 + \hat{\mathfrak{m}}$ we have $e^x e^y \in 1 + \hat{\mathfrak{m}}$ and since the exponential map is a bijection there is one and only one $z \in \hat{\mathfrak{m}}$ such that $e^z = e^x e^y$.

We have the relation

$$\begin{aligned} \Delta(e^z) = \Delta(e^x e^y) &= \Delta(e^x)\Delta(e^y) \\ &= (e^x \otimes e^x)(e^y \otimes e^y) \\ &= e^z \otimes e^z \,. \end{aligned}$$

Applying 7.3 we find that z is a primitive element and by $\widehat{7.1}$ $x \in \hat{L}_X$. q.e.d.

Now, let X be an arbitrary set and let $z(x,y)$ denote the element of $\hat{L}_{\{x,y\}} \subset \hat{L}_X$ such that $e^x e^y = e^{z(x,y)}$ for all $x, y \in X$.

We have $z(x,y) = \sum_{n=1}^{\infty} z_n(x,y)$ where $z_n(x,y) \in L_X^n$.

Explicitly, the values of the first three homogeneous components of $z(x,y)$ are

$$\begin{aligned} z_1(x,y) &= x+y \\ z_2(x,y) &= \tfrac{1}{2}[x,y] \\ z_3(x,y) &= \tfrac{1}{12}[x,[x,y]] + \tfrac{1}{12}[y,[y,x]] \end{aligned}$$

and it is clear that $z(x,0) = x$, $z(0,y) = y$, and $z(z(w,x),y) = z(w,z(x,y))$.

8. Explicit formula

Define a map $\Phi : \mathfrak{m} \to L_X$ ($\mathfrak{m} \subset \mathrm{Ass}_X$) as follows:

$$\Phi(x_1 \cdots x_n) = [x_1,[x_2,\ldots,[x_{n-1},x_n]\cdots] = \mathrm{ad}(x_1)\cdots\mathrm{ad}(x_{n-1})(x_n)$$

where $x_i \in X$.

Now define $\phi : \mathfrak{m} \to L_X$ by $\phi(x_1 \cdots x_n) = \frac{1}{n}\Phi(x_1 \cdots x_n)$.

Theorem 8.1. *The map ϕ is a retraction of $\mathfrak{m}$ onto L_X, i.e., $\phi|_{L_X} = \mathrm{id}_{L_X}$.*

Proof. We have to prove that $\Phi(u) = nu$ if $u \in L_X^n$.

Let $\theta : \mathrm{Ass}_X \to \mathrm{End}(L_X)$ be the algebra homomorphism which extends the Lie algebra homomorphism $\mathrm{ad} : L_X \to \mathrm{End}(L_X)$.

Lemma 8.2. *The relation $\Phi(uv) = \theta(u)\Phi(v)$ holds for $u \in \mathrm{Ass}_X$ and $v \in \mathfrak{m}$.*

Proof of Lemma. Since Φ and θ are linear it is enough to consider the case $u = x_1 \cdots x_n$, $x_i \in X$ and we proceed by induction on n.

If $n = 1$ then it is trivial.

Now suppose $n > 1$, then

$$\begin{aligned} \Phi(x_1 \cdots x_n v) &= \theta(x_1)\Phi(x_2 \cdots x_n v) = \theta(x_1)\theta(x_2 \cdots x_n)\Phi(v) \\ &= \theta(x_1 \cdots x_n)\Phi(v) . \end{aligned}$$

This concludes the proof of the lemma.

We now prove that $\Phi(u) = nu$ for $u \in L_X^n$ by induction on n.

If $n = 1$ the property is obvious.

Suppose $n > 1$, then $u = \sum[v_i, w_i]$ and it is enough to prove this when $u = [v,w]$ with $v \in L_X^p$, $w \in L_X^q$, $p + q = n$, $p,q > 0$.

Using the fact that $\theta(v) = \mathrm{ad}\, v$ and $\theta(w) = \mathrm{ad}\, w$ we get

$$\begin{aligned} \Phi([v,w]) &= \Phi(vw - wv) = \theta(v)\Phi(w) - \theta(w)\Phi(v) \\ &= q\theta(v)w - p\theta(w)v \\ &= q[v,w] - p[w,v] \\ &= (q+p)[v,w] = nu \end{aligned}$$

q.e.d.

Finally, we are prepared to give the explicit formula for $z(x,y) = \log(e^x e^y)$ for $x, y \in X$.

As before let us write $z = \sum_{n=1}^{\infty} z_n$ with $z_n \in L_X^n$.

Since $e^x e^y = \left(\sum_{p=0}^{\infty} \frac{x^p}{p!}\right)\left(\sum_{q=0}^{\infty} \frac{y^q}{q!}\right) = 1 + \sum_{p+q\geq 1} \frac{x^p y^q}{p!q!}$ we have

$$z = \log(e^x e^y) = \sum_{m=1}^{\infty} \frac{(-1)^{m+1}}{m} \left(\sum_{p+q\geq 1} \frac{x^p y^q}{p!q!} \right)^m ,$$

so we obtain

$$z = \sum_{p_i+q_i\geq 1} \frac{(-1)^{m+1}}{m} \frac{x^{p_1} y^{q_1} x^{p_2} y^{q_2} \ldots x^{p_m} y^{q_m}}{p_1!q_1! \cdots p_m!q_m!} .$$

Applying the homomorphism Φ to the monomials which appear in this sum we get

$$\Phi(x^{p_1} y^{q_1} \cdots x^{p_m} y^{q_m}) = \mathrm{ad}(x)^{p_1}\, \mathrm{ad}(y)^{q_1} \cdots \mathrm{ad}(x)^{p_m}\, \mathrm{ad}(y)^{q_m - 1}(y)$$

if $q_m \geq 1$, and:

$$\Phi(x^{p_1} y^{q_1} \cdots x^{p_m}) = \mathrm{ad}(x)^{p_1}\, \mathrm{ad}(y)^{q_1} \cdots \mathrm{ad}(x)^{p_m - 1}(x) , \qquad \text{if } q_m = 0.$$

Notice that this is zero if $q_m \geq 2$, or if $q_m = 0$, $p_m \geq 2$. Hence, the only possible non-zero terms are those where $q_m = 1$, or $p_m = 1$, $q_m = 0$.

Hence, using the identity $z_n = \phi(z_n)$, we obtain the *explicit Campbell-Hausdorff formula* (in Dynkin's form):

$$z_n = \frac{1}{n} \sum_{p+q=n} (z'_{p,q} + z''_{p,q}) ,$$

where

$$z'_{p,q} = \sum_{\substack{p_1+\cdots+p_m=p \\ q_1+\cdots+q_{m-1}=q-1 \\ p_i+q_i\geq 1 \\ p_m\geq 1}} \frac{(-1)^{m+1}}{m} \frac{\mathrm{ad}(x)^{p_1}\, \mathrm{ad}(y)^{q_1} \cdots \mathrm{ad}(x)^{p_m}(y)}{p_1!q_1! \cdots p_m!}$$

and

$$z''_{p,q} = \sum_{\substack{p_1+\cdots+p_{m-1}=p-1 \\ q_1+\cdots+q_{m-1}=q \\ p_i+q_i\geq 1}} \frac{(-1)^{m+1}}{m} \frac{\mathrm{ad}(x)^{p_1}\, \mathrm{ad}(y)^{q_1} \cdots \mathrm{ad}(y)^{q_{m-1}}(x)}{p_1!q_1! \cdots q_{m-1}!} .$$

Exercises

1. Let X be a finite set, with $\mathrm{Card}(X) = d$. Show that the number of elements of M_X of length n is equal to:

$$2^{n-1} d^n \frac{1 \cdot 3 \cdot 5 \cdots (2n-3)}{n!}$$

2. Show that $L_X^n = [X, L_X^{n-1}]$ for $n \geq 2$.

3. Show that the center of L_X is 0 if $\mathrm{Card}(X) \neq 1$, and that the center of $L_X / \sum_{n>p} L_X^n$ is equal to L_X^p.
4. Let X be a denumerable set with $\mathrm{Card}(X) \geq 2$, and let $\underline{H}$ the set of all Hall families in M_X. Show that $\mathrm{Card}(\underline{H}) = \mathrm{Card}(\mathbf{R})$.
5. Show that the homomorphism $\theta : F_X \to \widehat{\mathrm{Ass}}_X^*$ defined in IV.6 is injective.

Chapter V. Nilpotent and Solvable Lie Algebras

In this chapter k denotes a field, and in V.5, concerning the serious theorems on solvable Lie algebras, a field of characteristic 0. All Lie algebras and modules are finite dimensional over k.

1. Complements on $\mathfrak{g}$-modules

Let $\mathfrak{g}$ be a Lie algebras over k. A *$\mathfrak{g}$-module* is a vector space V over k together with a k-bilinear map $\mathfrak{g} \times V \to V$, denoted by $(x, v) \mapsto xv$, which satisfies the condition $[x, y]v = xyv - yxv$ for all $x \in \mathfrak{g}$, $y \in \mathfrak{g}$, $v \in V$. The corresponding Lie homomorphism $\varrho : \mathfrak{g} \to \mathrm{End}(V)$ is called a *linear representation of* $\mathfrak{g}$, and V is called the *space of the representation.*

An arbitrary vector space V can be made into a $\mathfrak{g}$ module by putting $xv = 0$ for all $v \in V$, $x \in \mathfrak{g}$. We then say that $\mathfrak{g}$ *acts trivially* on V. Whenever we view k as a $\mathfrak{g}$-module we understand it with trivial action unless the contrary is stated.

Let V_1 and V_2 be $\mathfrak{g}$-modules. The tensor product $V_1 \otimes V_2$ can be made into a $\mathfrak{g}$-module in a unique way such that the rule

$$x(v_1 \otimes v_2) = (xv_1) \otimes v_2 + v_1 \otimes (xv_2) \tag{1}$$

holds. This can be checked directly, or seen from the diagram

$$U\mathfrak{g} \xrightarrow{\Delta} U\mathfrak{g} \otimes U\mathfrak{g} \xrightarrow{\varrho_1 \otimes \varrho_2} \mathrm{End}\, V_1 \otimes \mathrm{End}\, V_2 \longrightarrow \mathrm{End}(V_1 \otimes V_2)\ ,$$

where Δ is the diagonal map. The action (1) is sometimes called the *diagonal action* of $\mathfrak{g}$ on $V_1 \otimes V_2$.

Similarly, the space of k-linear maps $\mathrm{Hom}_k(V_1, V_2)$ becomes a $\mathfrak{g}$-module if we put

$$(xf)(v_1) = x(f(v_1)) - f(xv_1)\ , \qquad \text{for } x \in \mathfrak{g},\ v_1 \in V_1. \tag{2}$$

More generally, given a finite family of $\mathfrak{g}$-modules V_i and V, we make the space of k-multilinear maps from $\prod_i V_i$ to V into a $\mathfrak{g}$-module in the corresponding way.

If V is a $\mathfrak{g}$-module, an element $v \in V$ is *$\mathfrak{g}$-invariant* if $xv = 0$ for all $x \in \mathfrak{g}$. This seemingly weird terminology comes from the corresponding group situation: $xv = 0$ is equivalent to $v = (1 + \varepsilon x)v$. The set of all $\mathfrak{g}$-invariant elements in V is a $\mathfrak{g}$-submodule of V, the largest submodule on which $\mathfrak{g}$ acts trivially.

Example 1. A k-linear map $f : V_1 \to V_2$ is $\mathfrak{g}$-invariant for the action of $\mathfrak{g}$ on $\mathrm{Hom}_k(V_1, V_2)$ described above if and only if $f(xv_1) = xf(v_1)$, that is, if and only if f is a *homomorphism of $\mathfrak{g}$-modules.*

Example 2. (Invariant bilinear forms). An invariant bilinear form

$$B : V_1 \times V_2 \to k$$

is one satisfying the identity

$$B(xv_1, v_2) + B(v_1, xv_2) = 0\ .$$

(For $g = 1 + \varepsilon x$ this means $B(gv_1, gv_2) = B(v_1, v_2)$.) Let V be a $\mathfrak{g}$-module and $\varrho : \mathfrak{g} \to \operatorname{End} V$ the corresponding linear representation. Define

$$B_\varrho(x, y) = \operatorname{Tr}_V\big(\varrho(x)\varrho(y)\big)\ ,$$

where $\operatorname{Tr}_V \alpha$ denotes the *trace* of a k-linear transformation $\alpha : V \to V$.

Proposition 1.1. *B_ϱ is a symmetric bilinear form on $\mathfrak{g} \times \mathfrak{g}$ which is $\mathfrak{g}$-invariant for the adjoint representation of $\mathfrak{g}$ on $\mathfrak{g}$.*

The rule $\operatorname{Tr}_V(\alpha\beta) = \operatorname{Tr}_V(\beta\alpha)$ shows the symmetry of B_ϱ. To prove its invariance we must show that the following expression

$$\operatorname{Tr}_V\big(\varrho([x, x_1])\varrho(x_2) + \varrho(x_1)\varrho([x, x_2])\big)$$
$$= \operatorname{Tr}_V\big(\varrho(x)\varrho(x_1)\varrho(x_2) - \varrho(x_1)\varrho(x)\varrho(x_2) + \varrho(x_1)\varrho(x)\varrho(x_2) - \varrho(x_1)\varrho(x_2)\varrho(x)\big)$$

is zero. To do so we cancel the middle two terms and use again the symmetry rule above with $\alpha = \varrho(x)$, and $\beta = \varrho(x_1)\varrho(x_2)$.

Definition 1.2. The *Killing form* is the invariant symmetric bilinear form $B(x, y) = \operatorname{Tr}(\operatorname{ad} x \operatorname{ad} y)$ on $\mathfrak{g}$ which is obtained by taking ϱ to be the adjoint representation in the preceding example.

2. Nilpotent Lie algebras

Let $\mathfrak{g}$ be a finite dimensional Lie algebra over a field k. The *descending central series* of ideals of $\mathfrak{g}$ is defined by $C^1\mathfrak{g} = \mathfrak{g}$ and $C^n\mathfrak{g} = [\mathfrak{g}, C^{n-1}\mathfrak{g}]$ for $n \geq 2$. (Here we write $[V, W]$ for the image of $V \otimes W$ under the map $(x \otimes y) \mapsto [x, y]$. We leave the proof of the rule $[C^r\mathfrak{g}, C^s\mathfrak{g}] \subset C^{r+s}\mathfrak{g}$ as an exercise for the reader.)

Theorem 2.1. *The following conditions are equivalent:*

(i) *There exists an integer n such that $C^n\mathfrak{g} = (0)$.*

(ii) *There exists an integer n such that*

$$[x_1, [x_2, [x_3, \ldots, x_n] \cdots]] = (\operatorname{ad} x_1)(\operatorname{ad} x_2) \cdots (\operatorname{ad} x_{n-1}) x_n = 0$$

for every n-tuple of elements (x_i) in $\mathfrak{g}$.

(iii) *$\mathfrak{g}$ is a succession of central extensions of abelian Lie algebras: that is, there exists a chain of ideals $\mathfrak{g} = \mathfrak{a}_1 \supset \mathfrak{a}_2 \supset \cdots \supset \mathfrak{a}_n = (0)$ such that $\mathfrak{a}_i/\mathfrak{a}_{i+1}$ is the center of $\mathfrak{g}/\mathfrak{a}_{i+1}$ for each i, or in other words, such that $[\mathfrak{g}, \mathfrak{a}_i] \subset \mathfrak{a}_{i+1}$ for all i.*

The proof in the form (i) $\Rightarrow$ (iii) $\Rightarrow$ (ii) $\Rightarrow$ (i) is completely trivial. Notice that the chain of ideals $C_n\mathfrak{g}$ is the most rapidly descending chain with the properties described in (iii). If $(\mathfrak{a}_i)$ is as in (iii), then $C^n\mathfrak{g} \subset \mathfrak{a}_n$ for all n.

Definition 2.2. If the conditions of Theorem 2.1 are satisfied, $\mathfrak{g}$ is called *nilpotent.*

Example. Let V be a vector space, and let $F = (V_i)$ be a *flag* in V, that is a sequence of subspaces $(0) = V_0 \subset V_1 \subset \cdots \subset V_n = V$ such that $\dim V_i = i$. Let $\mathfrak{u}(F) = \{ u \in \operatorname{End} V \mid uV_i \subset V_{i-1} \text{ for all } i \geq 1 \}$. Thus $\mathfrak{u}(F)$ is the set of endomorphisms of V which carry each V_i into itself *and* induce the zero endomorphism on the quotient space V_i/V_{i-1} for each $i \geq 1$. Obviously, $\mathfrak{u}(F)$ is an associative subalgebra of $\operatorname{End} V$, and *a fortiori* it is a Lie subalgebra under the bracket $[x, y] = xy - yx$. In terms of a basis (v_i) for V which is adapted to F in the sense that $V_i = kv_1 + \cdots + kv_i$, the algebra $\mathfrak{u}(F)$ consists of those endomorphisms whose matrix is *strictly superdiagonal*, that is, has zeros on and below the main diagonal. To show that $\mathfrak{u}(F)$ is nilpotent, define $\mathfrak{u}_k(F) = \{ u \in \operatorname{End}(V) \mid uV_i \subset V_{i-k} \}$ for all $i \geq k$, note that $\mathfrak{u}\mathfrak{u}_k \subset \mathfrak{u}_{k+1}$, and $\mathfrak{u}_k\mathfrak{u} \subset \mathfrak{u}_{k+1}$, hence $[\mathfrak{u}, \mathfrak{u}_k] \subset \mathfrak{u}_{k+1}$, hence $\mathfrak{g}$ is nilpotent because $\mathfrak{u}_k = 0$ for large k.

3. Main theorems

The following theorem offers some justification for the terminology "nilpotent":

Theorem 3.1. *$\mathfrak{g}$ is nilpotent if and only if $\operatorname{ad} x$ is nilpotent for each $x \in \mathfrak{g}$.*

We will at the same time consider:

Theorem 3.2 (Engel). *Let $\varrho : \mathfrak{g} \to \operatorname{End} V$ be a linear representation of $\mathfrak{g}$ on the vector space V such that $\varrho(x)$ is nilpotent for each $x \in \mathfrak{g}$. Then there exists a flag $F = (V_i)$ in V such that $\varrho(\mathfrak{g}) \subset \mathfrak{u}(F)$.*

The converse of Theorem 3.2 is trivial, because a strictly superdiagonal matrix is nilpotent. The meaning of Theorem 3.2 is that if for each individual $x \in \mathfrak{g}$ there exists a flag $F_X = \{V_{x,i}\}$ such that $\varrho(x)V_{x,i} \subset V_{x,(i-1)}$, then there exists one flag F which works for all x simultaneously. Theorem 3.2 is equivalent to:

Theorem 3.2$'$. *Under the hypotheses of Theorem 3.2, if $V \neq (0)$, then there exists a $v \in V$, $v \neq 0$, such that $\varrho(x)v = 0$ for all $x \in \mathfrak{g}$.*

Indeed, if Theorem 3.2$'$ holds, then Theorem 3.2 follows immediately by induction on $\dim V$. A flag $\bar{F}$ in $\bar{V} = V/kv$ lifts to a flag on V with the desired properties, of $\varrho(\mathfrak{g})v = 0$.

We shall now prove 3.2′ in seven steps:

Step 1: Since both the hypothesis and the conclusion concern only the image $\varrho(\mathfrak{g})$ in End V, we can replace $\mathfrak{g}$ by its image and assume $\mathfrak{g} \subset \operatorname{End} V$.

Step 2: Then ad x is nilpotent for each $x \in \mathfrak{g}$. Namely, we can write

$$(\operatorname{ad} x)y = L_x y - R_x y \ ,$$

where L_x and R_x are the k-linear endomorphisms of End V defined by $\alpha \mapsto x\alpha$, and $\alpha \mapsto \alpha x$, respectively. But L_x and R_x are nilpotent by hypothesis and commute. Hence $L_x - R_x$ is nilpotent. (Show that if $\alpha^m = 0$, $\beta^n = 0$ and $\alpha\beta = \beta\alpha$, then $(\alpha - \beta)^{m+n-1} = 0$.)

Step 3: By induction on dim $\mathfrak{g}$, we may assume 3.2′ holds for all Lie algebras $\mathfrak{h}$ such that $\dim \mathfrak{h} < \dim \mathfrak{g}$.

Step 4: Let $\mathfrak{h} \subset \mathfrak{g}$ be a Lie subalgebra, $\mathfrak{h} \neq \mathfrak{g}$. Let $\mathfrak{u} = \{ x \in \mathfrak{g} \mid (\operatorname{ad} x)\mathfrak{h} \subset \mathfrak{h} \}$ be the *normalizer of* $\mathfrak{h}$ *in* $\mathfrak{g}$, that is, the largest subalgebra of $\mathfrak{g}$ in which $\mathfrak{h}$ is an ideal. Our aim is to prove $\mathfrak{a}$ is *strictly larger than* $\mathfrak{h}$. (The reader familiar with the theory of p-groups will note the analogy.) The Lie algebra $\mathfrak{h}$ operates on $\mathfrak{h}$ and on $\mathfrak{g}/\mathfrak{h}$ through nilpotent maps. Since $\dim \mathfrak{h} < \dim \mathfrak{g}$, there is a non-zero vector $\bar{x} = x + \mathfrak{h}$ in $\mathfrak{g}/\mathfrak{h}$ invariant (i.e. killed) by $\mathfrak{h}$. For $y \in \mathfrak{h}$ we have then $(\operatorname{ad} x)y = -(\operatorname{ad} y)x \in \mathfrak{h}$ because $(\operatorname{ad} y)\bar{x} = 0$. Thus $x \in \mathfrak{u}$ an our claim is proved.

Step 5: If $\mathfrak{g} \neq (0)$, there exists an ideal $\mathfrak{h}$ in $\mathfrak{g}$ of codimension 1. Indeed, let $\mathfrak{h}$ be a maximal Lie subalgebra of $\mathfrak{g}$ different from $\mathfrak{g}$. Then, by step 4, the normalizer of $\mathfrak{h}$ is all of $\mathfrak{g}$, that is, $\mathfrak{h}$ is an ideal in $\mathfrak{g}$. The inverse image in $\mathfrak{g}$ of a line in $\mathfrak{g}/\mathfrak{h}$ is a subalgebra of $\mathfrak{g}$ strictly bigger than $\mathfrak{h}$, hence is all of $\mathfrak{g}$, and $\mathfrak{g}/\mathfrak{h}$ is therefore one dimensional.

We now choose such an ideal $\mathfrak{h}$.

Step 6: Let $W = \{ v \in V \mid \mathfrak{h}v = 0 \}$. Then W is stable by $\mathfrak{g}$. This depends only on the fact that $\mathfrak{h}$ is an ideal. For $x \in \mathfrak{g}$, $y \in \mathfrak{h}$ we have $yxv = xyv - [x, y]v = 0$ if $v \in W$.

Step 7: $W \neq (0)$ by induction ($\dim \mathfrak{h} < \dim \mathfrak{g}$). Take $y \in \mathfrak{g}$, $y \notin \mathfrak{h}$. Since y is nilpotent, y kills some non-zero element in W. This element is then killed by $\mathfrak{g} = \mathfrak{h} + ky$. q.e.d.

We now prove Theorem 3.1. If $\mathfrak{g}$ is nilpotent then ad x is nilpotent for each $x \in \mathfrak{g}$ by condition (ii) of Theorem 2.1. Conversely, if ad x is nilpotent for all x, then, applying Engel's theorem to the adjoint representation, we see that there exists a flag $(0) \subset \mathfrak{a}_1 \subset \mathfrak{a}_2 \subset \cdots \subset \mathfrak{a}_n = \mathfrak{g}$ of subspaces of $\mathfrak{g}$, such that $[\mathfrak{g}, \mathfrak{a}_i] \subset \mathfrak{a}_{i-1}$ for all i, and consequently $\mathfrak{g}$ is nilpotent by criterion (iii) of Theorem 2.1.

3*. The group-theoretic analog of Engel's theorem

Let V be a finite dimensional vector space over k. An element $g \in \mathrm{GL}(V)$ is called *unipotent* if it satisfies one and hence all of the following three conditions, whose equivalence we leave as an exercise for the reader:

(i) $g = 1 + n$ with n nilpotent.

(ii) In suitable coordinates g is represented by a matrix having 1's on the main diagonal and zero below.

(iii) All eigenvalues of g are 1.

Theorem (Kolchin). *Let G be a subgroup of $\mathrm{GL}(V)$ such that each element $g \in G$ is unipotent. Then there exists a flag $F = \{V_i\}$ in V such that $GV_i = V_i$ for all i.*

In other words, there is a coordinate system in which all elements $g \in G$ are represented simultaneously by triangular matrices (and hence by triangular matrices with 1's on the diagonal since the eigenvalues are all 1 by hypothesis).

The theorem will follow by induction on $\dim V$ if we can show that, under the given hypothesis, if $V \neq (0)$ there exists a $v \in V$, $v \neq 0$, which is left fixed by G. The equations $(g-1)v = 0$, for $g \in G$, are linear, and will therefore have a non-trivial solution v over k if and only if they have one over the algebraic closure $\bar{k}$ of k, i.e., in $V \otimes_k \bar{k}$. We may therefore suppose that k is algebraically closed. Furthermore, replacing V by a G-submodule we may suppose that V is a *simple* G-module. From the density theorem, or Burnside's theorem (Bourbaki, *Alg.*, Ch 8, §4, n° 3) it follows that the elements of G span all of $\mathrm{End}_k(V)$ linearly, because $\sum_{g \in G} kg$ is a k-subalgebra of $\mathrm{End}_k(V)$.

On the other hand, for each $g = 1 + n \in G$ we have

$$\mathrm{Tr}_V(g) = \mathrm{Tr}_V(1) + \mathrm{Tr}_V(n) = \mathrm{Tr}_V(1)$$

because nilpotent endomorphisms have zero eigenvalues hence zero trace. Thus, $\mathrm{Tr}_V(g)$ *is independent of* $g \in G$, and for every $g' \in G$ we have $\mathrm{Tr}(ng') = \mathrm{Tr}((g-1)g') = \mathrm{Tr}(gg' - g') = \mathrm{Tr}(gg') - \mathrm{Tr}(g') = 0$. Since the g' span $\mathrm{End}_k(V)$ it follows that $\mathrm{Tr}_V(n\alpha) = 0$ for all $\alpha \in \mathrm{End}_k(V)$, and consequently $n = 0$, i.e., $g = 1$. This is what we were trying to prove: G acts trivially on V.

4. Solvable Lie algebras

The *derived series* $(D^n\mathfrak{g})$ of ideals in $\mathfrak{g}$ is defined inductively by $D^1\mathfrak{g} = \mathfrak{g}$, and $D^n\mathfrak{g} = [D^{n-1}\mathfrak{g}, D^{n-1}\mathfrak{g}]$ for $n > 1$.

Theorem 4.1. *The following conditions are equivalent:*

(i) *There exists an integer n such that $D^n\mathfrak{g} = (0)$.*

(ii) *There exists an integer n such that for every family of 2^n elements $x_\nu \in \mathfrak{g}$ we have*

$$[[[\cdots][\quad]][[\quad][\quad]]] = 0\,.$$

(iii) $\mathfrak{g}$ *is a successive extension of abelian Lie algebras, that is, there exists a sequence of ideals $\mathfrak{g} = \mathfrak{a}_1 \supset \mathfrak{a}_2 \supset \cdots \supset \mathfrak{a}_n = (0)$ such that $\mathfrak{a}_i/\mathfrak{a}_{i+1}$ is abelian, i.e., $[\mathfrak{a}_i, \mathfrak{a}_i] \subset \mathfrak{a}_{i+1}$, for all i.*

It is trivial that (i) $\Rightarrow$ (iii) $\Rightarrow$ (ii) $\Rightarrow$ (i).

Definition 4.2. If $\mathfrak{g}$ satisfies the three equivalent conditions of the preceding theorem, $\mathfrak{g}$ is said to be a *solvable Lie algebra.*

Example. Let $F = (V_i)$ be a flag in a finite dimensional vector space V. Let $\mathfrak{b}(F) = \{\, x \in \operatorname{End} V \mid xV_i \subset V_i \text{ for all } i \,\}$. If we adapt the coordinate system to the flag, then $\mathfrak{b}(F)$ consists of the triangular matrices. It is easy to check that $\mathfrak{b}(F)/\mathfrak{a}(F)$ is abelian, and consequently $\mathfrak{b}(F)$ is solvable.

5. Main theorem

In this section our field k is of characteristic 0. The main theorem on solvable Lie algebras is:

Theorem 5.1 (Lie). *Let $\mathfrak{g}$ be a solvable Lie algebra over an algebraically closed field k of characteristic 0. Let ϱ be a linear representation of $\mathfrak{g}$ with representation space V. Then there exists a flag $F = (V_i)$ in V such that $\varrho(\mathfrak{g}) \subset \mathfrak{b}(F)$.*

This reduces, by induction on $\dim V$, to:

Theorem 5.1′. *Under the hypotheses of Theorem 5.1, if $V \neq (0)$, there exists $v \in V$, $v \neq 0$, such that v is an eigenvector for $\varrho(x)$ for all $x \in \mathfrak{g}$.*

Note that if v is such an eigenvector, it determines a map $\chi : \mathfrak{g} \to k$ such that $\varrho(x)v = \chi(x)v$ for all $x \in \mathfrak{g}$.

Main Lemma. *Let $\mathfrak{g}$ be a Lie algebra, over a field k of characteristic 0, $\mathfrak{h}$ an ideal in $\mathfrak{g}$, V a $\mathfrak{g}$-module, $v \in V$, $v \neq 0$, $\chi : \mathfrak{h} \to k$ such that $hv = \chi(h)v$ for all $h \in \mathfrak{h}$. Then $\chi([x,h]) = 0$ for $x \in \mathfrak{g}$, $h \in \mathfrak{h}$.*

Take $x \in \mathfrak{g}$. Let V_i be the subspace of V generated by the vectors v, xv, $\ldots$, $x^{i-1}v$. Thus $(0) = V_0 \subset V_1 \subset \cdots \subset V_i \subset V_{i+1}$. Let n be minimal > 0 such that $V_n = V_{n+1}$. Then $\dim V = n$, and $xV_n \subset V_n$, and $V_n = V_{n+k}$ for all $k \geq 0$. Claim: For $h \in \mathfrak{h}$, $hx^iv \equiv \chi(h)x^iv \pmod{V_i}$, for all $i \geq 0$. We prove this by induction on i. For $i = 0$ this is the definition of χ. For $i > 0$,

$$hx^iv = hxx^{i-1}v = xhx^{i-1}v - [x,h]x^{i-1}v\,.$$

Writing $hx^{i-1}v = \chi(h)x^{i-1}v + v'$ with $v' \in V_{i-1}$, and using $xV_{i-1} \subset V_i$, and $\mathfrak{h}V_i \subset V_i$, we are done. It follows that, with respect to the basis v, xv, ..., $x^{n-1}v$, the endomorphism of V_n produced by an element $h \in \mathfrak{h}$ is represented by a triangular matrix with diagonal entries $\chi(h)$. Thus, $\mathrm{Tr}_{V_n}(h) = n\chi(h)$. Replacing h by $[x,h]$ we conclude $n\chi([x,h]) = 0$, because $\mathrm{Tr}_{V_n}([x,h]) = \mathrm{Tr}_{V_n}(xh - hx) = 0$ (notice that $xV_n \subset V_n$).

Using the lemma, we prove Theorem 5.1′ by induction on $\dim \mathfrak{g}$. If $\dim \mathfrak{g} = 0$ the statement is trivial. Assume $\dim \mathfrak{g} > 0$. Then, since $\mathfrak{g}$ is solvable, $D\mathfrak{g} = [\mathfrak{g}, \mathfrak{g}] \neq \mathfrak{g}$. Let $\mathfrak{h}$ be a subspace of $\mathfrak{g}$ of codimension 1, containing $D\mathfrak{g}$. Then $\mathfrak{h}/D\mathfrak{g}$ is an ideal in $\mathfrak{g}/D\mathfrak{g}$ because the latter is abelian, and consequently $\mathfrak{h}$ is an ideal in $\mathfrak{g}$. By induction there is $v \in V$, $v \neq 0$, and $\chi : \mathfrak{h} \to k$ such that $hv = \chi(h)v$ for all $h \in \mathfrak{h}$. Let $W = \{ w \in V \mid hw = \chi(h)w \text{ for all } h \in \mathfrak{h} \}$. By construction, W is a non-zero linear subspace of V, and using the main lemma we can show that W is stable under $\mathfrak{g}$. If $w \in W$, $x \in \mathfrak{g}$, then for $h \in \mathfrak{h}$,

$$hxw = xhw - [x,h]w = \chi(h)xw - \chi([x,h])w ,$$

and since the last term is zero, it follows that $xw \in W$.

Now let $x \in \mathfrak{g}$, $x \notin \mathfrak{h}$. Since x maps W into W, and k is algebraically closed, there is an eigenvector $v_0 \in W$ for x. This v_0 is an eigenvector for $kx + \mathfrak{h} = \mathfrak{g}$. q.e.d.

To see that the theorem is false in characteristic $\neq 0$, consider the Lie algebra $\mathfrak{sl}_2$ of 2×2 matrices with trace 0 in characteristic 2. It is nilpotent of dimension 3, but in its standard representation on the space of column vectors of length 2, there is no eigenvector.

We close this section with two corollaries of Lie's Theorem.

Corollary 5.2. *If $\mathfrak{g}$ is a solvable Lie algebra over an algebraically closed field of characteristic zero, then there exists a flag of ideals in $\mathfrak{g}$.*

We need only apply Lie's theorem to the adjoint representation.

Corollary 5.3. *If $\mathfrak{g}$ is solvable and k of characteristic zero, then $[\mathfrak{g}, \mathfrak{g}]$ is nilpotent.*

Since the statement is *linear*, we may suppose that k is algebraically closed. (If k' is an extension field of k, and $\mathfrak{g}' = \mathfrak{g} \otimes_k k'$, then it is obvious that $\mathfrak{g}$ is solvable (resp. nilpotent) if and only if $\mathfrak{g}'$ is solvable (resp. nilpotent), that $[\mathfrak{g}, \mathfrak{g}]' = [\mathfrak{g}', \mathfrak{g}']$, etc.) By the preceding corollary there is a flag $(\mathfrak{g}_i)$ of ideals in $\mathfrak{g}$, say $\mathfrak{g} \supset \mathfrak{g}_1 \supset \mathfrak{g}_2 \supset \cdots \supset \mathfrak{g}_n = 0$. Let $x \in [\mathfrak{g}, \mathfrak{g}]$. Then $\mathrm{ad}\, x\, \mathfrak{g}_i \subset \mathfrak{g}_{i+1}$ because $\mathrm{End}(\mathfrak{g}_i/\mathfrak{g}_{i+1}) \simeq k$ is commutative. Hence $\mathrm{ad}\, x$ is nilpotent on $\mathfrak{g}$, and all the more so on $[\mathfrak{g}, \mathfrak{g}]$. By Theorem 3.1 we conclude that $[\mathfrak{g}, \mathfrak{g}]$ is nilpotent.

Remark. Conversely, if $[\mathfrak{g}, \mathfrak{g}]$ is nilpotent, it is clear that $\mathfrak{g}$ is solvable.

5*. The group theoretic analog of Lie's theorem

A group G is called *solvable* if it can be obtained by a finite sequence of extensions of abelian groups. One defines a sequence of subgroups $G^{(n)}$ of G by $G^{(1)} = G$, and $G^{(n)} = (G^{(n-1)}, G^{(n-1)})$ for $n > 1$. Then the solvability of G is equivalent to $G^{(n)} = (1)$ for some n.

Let G be a topological group and $\varrho : G \to \mathrm{GL}(V)$ a continuous homomorphism of G into the group of automorphisms of a finite dimensional vector space V over $\mathbf{C}$.

Theorem 5.1*. *If G is solvable and connected, there exists a flag F in V which is invariant by $\varrho(x)$ for all $x \in G$.*

The representation ϱ is called *irreducible* if $V \neq (0)$ and if V and (0) are the only subspaces of V which are invariant by $\varrho(x)$ for all $x \in G$. Theorem 5.1* implies obviously

Corollary 5.2*. *If G is solvable and connected, and the representation ϱ is irreducible, then* $\dim V = 1$.

Conversely, by induction on $\dim V$, this corollary trivially implies the theorem.

Corollary 5.3*. *A compact solvable connected topological group is abelian.*

By the Peter-Weyl theorem, for any compact group G there exists a family of irreducible representations $\varrho_\alpha : G \to \mathrm{GL}(V_\alpha)$ such that the map $G \to \prod_\alpha \mathrm{GL}(V_\alpha)$ is injective. If $\dim V_\alpha = 1$ for each α, it follows that G is abelian.

In proving the theorem we will use the following terminology: an element $v \in V$ is an *eigenvector* for a subgroup $H \subset G$ if $v \neq 0$ and if $hv \in \mathbf{C}v$ for all $h \in H$. An eigenvector v for H defines a *character* $\chi_v : H \to \mathbf{C}^*$ such that $\varrho(h)v = \chi_v(h)v$ for all $h \in H$. Of course the function χ_v is continuous because ϱ is. The number of distinct characters of H arising from eigenvectors $v \in V$ is finite, and in fact $\leq \dim V$. Indeed, suppose that $v_1, \ldots, v_r$ is a maximal independent set of eigenvectors for H in V, and let $\chi_1, \ldots, \chi_r$ be the corresponding characters. Then if v is an eigenvector with character χ we have $v = \sum a_i v_i$ with $a_i \in \mathbf{C}$, and applying $\varrho(h)$ we find $a_i \chi(h) = a_i \chi_i(h)$ for each i. Hence $\chi = \chi_i$ for some i, because not all a_i are zero.

Main Lemma*. *Suppose that G is connected. Let v be an eigenvector for a normal subgroup H. Then $\chi_v(x^{-1}hx) = \chi_v(h)$ for all $x \in G$ and $h \in H$.*

Notice the analogy with the main lemma of the preceding section. A simple computation shows that $\chi_v(x^{-1}hx) = \chi_{xv}(h)$. As remarked above, there are only a finite number of characters of H of the form χ_{xv} for $x \in G$. Hence, the subgroup $S = \{ x \in G \mid \chi_{xv} = \chi_v \}$ is of finite index in G. But S is closed in G, since it is the set of common zeros of the continuous functions $x \mapsto \chi_v(x^{-1}hx) - \chi_v(h)$ as h ranges over H. The decomposition of G into cosets of S is an expression of G as the disjoint union of a finite number of non-empty closed sets. Since G is connected, it follows that $S = G$ which is what we want.

We now prove the theorem by induction on the smallest number n such that $G^{(n)} = 1$. If $n = 1$ then $G = (1)$ and the theorem is trivial. Suppose therefore that $n > 1$, so that $G^{(2)} \neq G^{(1)} = G$. By induction, we can assume the theorem is true for $G^{(2)}$, because $G^{(2)}$ is connected. Indeed, let C be the set of commutators in G. As image of $G \times G$ under the map $x \times y \mapsto xyx^{-1}y^{-1}$, C is connected. Let C^m denote the set of elements of G which can be expressed as the product of m elements in C. The set C^m is connected because it is the continuous image of $C \times C \times \cdots \times C$ (m times). Since $u \in C$ implies $u^{-1} \in C$, the subgroup $G^{(2)}$ generated by C is the union of the connected sets C^m, and is connected because the C^m have a common point, namely 1.

By induction, there exists an eigenvector v_0 for $G^{(2)}$ in V. Let

$$\chi_0 : G^{(2)} \to \mathbf{C}^*$$

be the corresponding character. By the main lemma, the set of all $v \in V$ such that $\varrho(h)v = \chi_0(h)v$ for all $h \in G^{(2)}$ is invariant under $\varrho(G)$.

Suppose ϱ irreducible. It follows that $\varrho(h)v = \chi(h)v$ for all $v \in V, h \in G^{(2)}$. Now let $x \in G$. Let H be the subgroup of G generated by x and $G^{(2)}$. Since $H \supset G^{(2)}$, H is normal in G. Since $\mathbf{C}$ is algebraically closed, there exists an eigenvector v_1 for the operator $\varrho(x)$. By the above, v_1 is an eigenvector for $G^{(2)}$, and hence for H. Let $\chi_1 : H \to \mathbf{C}^*$ be the corresponding character. By the main lemma again, the set of all $v \in V$ such that $\varrho(h)v = \chi_1(h)v$ for all $h \in H$ is invariant under $\varrho(G)$ and hence is all of V. Hence, in particular $\varrho(x)v \in \mathbf{C}v$ for each $v \in V$. Since x was arbitrary in G, we conclude that $\dim V = 1$. Thus Corollary 5.2* and Theorem 5.1* are proved.

Remark. In fact, Lie's theorem and its group theoretic analog imply each other directly. Granting the group statement, we get the Lie algebra statement in case $k = \mathbf{C}$, by considering the connected Lie group attached to a given Lie algebra. The case of an arbitrary algebraically closed k of characteristic zero is reduced to $k = \mathbf{C}$ by the Lefschetz principle: Take a finitely generated subfield k' of k containing the structure constants for $\mathfrak{g}$ and for the action $\mathfrak{g} \times V \to V$, and imbed k' in $\mathbf{C}$. The descent from $\mathbf{C}$ to $\overline{k'}$ is easy.

Conversely, if we grant Lie's theorem we can get the group statement by considering the closure of $\varrho(G)$ in $\mathrm{GL}(V)$ as a real Lie group and applying Lie's theorem to its Lie algebra.

6. Lemmas on endomorphisms

Let k be an algebraically closed field of characteristic zero, and let V be a finite dimensional vector space over k. An element $u \in \operatorname{End} V$ is called *semisimple* if its eigenvectors span V or, what is the same, if there exists a system of coordinates in which it is represented by a diagonal matrix.

Lemma 6.1. *For each $u \in \operatorname{End} V$ there exist a semisimple s and a nilpotent n in $\operatorname{End} V$ such that $sn = ns$ and $u = s + n$, and s and n are uniquely determined by these conditions. Moreover, there exist polynomials S and N (depending on u) such that $S(0) = 0 = N(0)$ and $s = S(u)$ and $n = N(u)$.*

Let $\det(T - u) = \prod(T - \lambda_i)^{m_i}$ be the factorization of the characteristic polynomial of u into a product of powers of *distinct* linear factors $T - \lambda_i$. For each i, let V_i be the kernel of the endomorphism $(u - \lambda_i)^{m_i} : V \to V$. Then $V = \bigoplus V_i$ (direct sum), $\dim V_i = m_i$, and $uV_i \subset V_i$. Suppose $u = s + n$ is a solution to our problem. Since s commutes with n, it commutes with u, hence with $(u - \lambda_i)^{m_i}$. Therefore $sV_i \subset V_i$ for each i. Since $u - s$ is nilpotent, the eigenvalues of s on V_i are the same as those of u. But by construction, u has the unique eigenvalue λ_i on V_i. Since s is semisimple, it follows that the restriction of s to V_i is scalar multiplication by λ_i. On the other hand, taking this as definition of s, and putting $n = u - s$ (so that on V_i, n_i has the same effect as $u - \lambda_i$) we obviously obtain a solution to the problem. Let $S(T)$ be a polynomial satisfying

$$S(T) \equiv \lambda_i \pmod{(T - \lambda_i)^{m_i}}, \quad \text{and} \quad S(T) \equiv 0 \pmod{T} .$$

(note the consistency of these two conditions in case $\lambda_i = 0$ for some i) and put $N(T) = T - S(T)$. Then $S(0) = 0 = N(0)$ and $s = S(u)$, $n = N(u)$ as required.

Consequence 6.2. *Let $u = s + n$ as in the preceding lemma. Suppose A and B are subspaces of V such that $A \subset B$ and $uB \subset A$. Then $sB \subset A$ and $nB \subset A$.*

Indeed if $P(T)$ is any polynomial in T without constant term, then $P(u)B \subset A$.

Let now $V^* = \operatorname{Hom}_k(V, k)$ be the dual of V, and for $p, q \geq 0$ let

$$V_{p,q} = \underbrace{V \otimes \cdots \otimes V}_{p-\text{times}} \otimes \underbrace{V^* \otimes \cdots \otimes V^*}_{q-\text{times}} .$$

We view $V_{p,q}$ as a module for the Lie algebra $\operatorname{End} V$ by means of the *diagonal action* discussed in §1. For $u \in \operatorname{End} V$, we let $u_{p,q}$ denote the corresponding endomorphism of $V_{p,q}$. For example,

$$u_{12} = u \otimes 1 \otimes 1 - 1 \otimes u^* \otimes 1 - 1 \otimes 1 \otimes u^* ,$$

where $u^* \in \operatorname{End} V^*$ is the "transpose" of u, defined by $\langle u^*y, x\rangle = \langle y, ux\rangle$, if we write $\langle y, x\rangle$ instead of $y(x)$ for $y \in V^*$, $x \in V$.

An important special case is that in which $p = q = 1$. There is a canonical isomorphism $V_{1,1} \xrightarrow{\sim} \operatorname{End} V$ which associates with $x \otimes y$ the endomorphism $x' \mapsto x\langle y, x'\rangle$. A simple computation shows that under this isomorphism, the element $u_{1,1} \in \operatorname{End} V_{1,1}$ corresponds to $\operatorname{ad} u \in \operatorname{End}(\operatorname{End} V)$.

Lemma 6.3. *If $u = s + n$ is the canonical decomposition of u as in Lemma 6.1, then $u_{p,q} = s_{p,q} + n_{p,q}$ is the canonical decomposition of $u_{p,q}$ for each p, q.*

We have $[s_{p,q}, n_{p,q}] = [s, n]_{p,q} = 0_{p,q} = 0$, hence $s_{p,q}$ and $n_{p,q}$ commute. If (x_i) is a basis for V consisting of eigenvectors of s, then the dual basis (x_i^*) of V^* consists of eigenvectors of s^*, and the basis $(x_{i_1} \otimes \cdots \otimes x_{i_p} \otimes x_{j_1}^* \otimes \cdots \otimes x_{j_q}^*)$ of $V_{p,q}$ consists of eigenvectors for $s_{p,q}$; hence $s_{p,q}$ is semisimple. The endomorphism $n_{p,q}$ is a sum of endomorphisms of the form $1 \otimes \cdots \otimes n \otimes \cdots \otimes 1$ or $1 \otimes \cdots \otimes n^* \otimes \cdots \otimes 1$, each of which is nilpotent, and which commute with each other; hence $n_{p,q}$ is nilpotent. We have $u_{p,q} = s_{p,q} + n_{p,q}$ because the map $u \mapsto u_{p,q}$ is linear. The lemma now results from the uniqueness of the canonical decomposition.

Let s be a semisimple element of $\operatorname{End} V$, and let $V = \bigoplus V_i$ be the corresponding direct decomposition, with $s|_{V_i} = \lambda_i$. Let $\phi : k \to k$ be a $\mathbf{Q}$-linear map.

Definition 6.4. $\phi(s)$ is the semisimple endomorphism of V such that $\phi(s)|_{V_i} = \phi(\lambda_i)$.

Thus, if s is represented by a diagonal matrix, the matrix representing $\phi(s)$ is obtained by applying ϕ to the entries of the matrix representing s. There is a polynomial $P(T)$ (depending on ϕ and s) such that $P(0) = 0$ and $P(s) = \phi(s)$. We need only solve the interpolation problem $P(\lambda_i) = \phi(\lambda_i)$, for each i, and $P(0) = 0$. So far we have only used the fact that ϕ maps k into k and $\phi(0) = 0$. The linearity of ϕ is needed to prove:

Lemma 6.5. *We have $\left(\phi(s)\right)_{p,q} = \phi(s_{p,q})$ for each p, q.*

The space $V_{p,q}$ is a direct sum of subspaces, the typical one of which is $V_{i_1} \otimes \cdots \otimes V_{i_p} \otimes V_{j_1}^* \otimes \cdots \otimes V_{j_q}^*$. On that subspace:

$s_{p,q}$ does scalar multiplication by $\lambda_{i_1} + \cdots + \lambda_{i_p} - \lambda_{j_1} - \cdots - \lambda_{j_q}$,

$\phi(s_{p,q})$ " " " by $\phi(\lambda_{i_1} + \cdots + \lambda_{i_p} - \lambda_{j_1} - \cdots - \lambda_{j_q})$,

and $\left(\phi(s)\right)_{p,q}$ " " " by $\phi(\lambda_{i_1}) + \cdots + \phi(\lambda_{i_p}) - \phi(\lambda_{j_1}) - \cdots - \phi(\lambda_{j_q})$.

Consequence 6.6. *Suppose $u = s + n$ is the canonical decomposition of $u \in \operatorname{End} V$. Suppose A and B are subspaces of $V_{p,q}$ such that $A \subset B$ and $u_{p,q}B \subset A$. Then for each $\mathbf{Q}$-linear map $\phi : k \to k$ we have $\phi(s)_{p,q}B \subset A$.*

By Lemma 6.3 and Consequence 6.2, we have $s_{p,q}B \subset A$. The result follows now because $\phi(s)_{p,q} = \phi(s_{p,q})$ is a polynomial in $s_{p,q}$ without constant term, as discussed in the remarks preceding Lemma 6.5.

Lemma 6.7. *Let $u = s + n$ as in Lemma 6.1. If $\mathrm{Tr}(u\phi(s)) = 0$ for all $\phi \in \mathrm{Hom}_{\mathbf{Q}}(k, k)$, then u is nilpotent.*

With notation as in the proof of 6.1, we have $\mathrm{Tr}(u\phi(s)) = \sum m_i\lambda_i\phi(\lambda_i) = 0$ for all $\phi \in \mathrm{Hom}_{\mathbf{Q}}(k, k)$. For those ϕ with $\phi(k) \subset \mathbf{Q}$ we can apply ϕ again, getting $0 = \sum m_i(\phi(\lambda_i))^2$, and consequently $\phi(\lambda_i) = 0$ for each i. Since this last holds for every $\phi \in \mathrm{Hom}(k, \mathbf{Q})$ it follows that $\lambda_i = 0$ for each i, that is, $s = 0$, and $u = n$ as contended. [Variant, remarked by Bergman: If $k = \mathbf{C}$, we need only assume $\mathrm{Tr}(u\phi(s)) = 0$ for one single ϕ, namely the complex conjugation map.]

The endomorphisms $\phi(s)$ are called *replicas* of s by Chevalley. We leave as an exercise the following characterization:

Theorem 6.8. *Let s and s' be semisimple elements of* End V. *Then s' is a replica of s* (i.e., there exists a ϕ such that $s' = \phi(s)$) *if and only if, for every p, q, every element of $V_{p,q}$ which is killed by s is also killed by s'.*

There is another characterization in terms of algebraic groups which is even nicer: Let $\mathfrak{g}$ be the set of replicas of s. Then it can be shown that $\mathfrak{g}$ is the Lie algebra of the smallest *algebraic* subgroup G of $\mathrm{GL}(V)$ whose Lie algebra contains s. Indeed the group G, or more properly, the group $G(k)$ of points of G with coordinates in k, consists of the automorphisms x of V such that, for each i, $x|_{V_i}$ is multiplication by a scalar $x_i \in k^*$, these scalars being subject to the relation $\prod x_i^{n_i} = 1$ for every vector $(\ldots, n_i, \ldots)$ of integers n_i such that $\sum n_i\lambda_i = 0$. [Cf. C. Chevalley, *Théorie des Groupes de Lie*, Tome II, *Groupes algébriques*, Ch II, §13–14, Hermann, Paris, 1951. Also *Algebraic Lie Algebras*, Annals of Math., Vol. 48, 1947, p. 91–100.]

7. Cartan's criterion

The following criterion for solvability is useful.

Theorem 7.1. *Let k be a field of characteristic zero, V a finite dimensional vector space over k, and $\mathfrak{g}$ a Lie subalgebra of* End V. *Then the following conditions are equivalent:*

(i) *$\mathfrak{g}$ is solvable.*

(ii) $\mathrm{Tr}(xy) = 0$ *for every $x \in \mathfrak{g}$ and $y \in D\mathfrak{g} = [\mathfrak{g}, \mathfrak{g}]$.*

Note first that the statement is linear, so that we can assume that k is algebraically closed (see the discussion after Corollary 5.3; by the "Lefschetz principle", that is, by choosing a finitely generated subfield $k' \subset k$ over which

V and $\mathfrak{g}$ are defined and imbedding k' in $\mathbf{C}$, we could even reduce to the case $k = \mathbf{C}$.)

(i) $\Rightarrow$ (ii) By Lie's theorem we can choose a flag (V_i) in V stable by $\mathfrak{g}$. Then $\mathrm{Tr}_V(xy) = \sum_i \mathrm{Tr}_{V_i/V_{i+1}}(xy) = 0$ because an element $y \in D\mathfrak{g}$ annihilates the one-dimensional $\mathfrak{g}$-modules V_i/V_{i+1}.

(ii) $\Rightarrow$ (i). Let $u \in D\mathfrak{g}$. By Engel's theorem, it suffices to show that u is nilpotent. Write $u = s + n$ as in Lemma 6.1. Then, by Lemma 6.7, it suffices to prove $\mathrm{Tr}\big(u\phi(s)\big) = 0$ for all $\phi \in \mathrm{Hom}_{\mathbf{Q}}(k,k)$. The problem is that $\phi(s)$ need not belong to $\mathfrak{g}$. Write $u = \sum c_\alpha[x_\alpha, y_\alpha]$, with $c_\alpha \in k$, and $x_\alpha, y_\alpha \in \mathfrak{g}$. Using the rule $\mathrm{Tr}([a,b]c) = \mathrm{Tr}(b[c,a])$, we have

$$\mathrm{Tr}\big(u\phi(s)\big) = \sum c_\alpha \,\mathrm{Tr}\big([x_\alpha, y_\alpha]\phi(s)\big) = \sum c_\alpha \,\mathrm{Tr}\big(y_\alpha[\phi(s), x_\alpha]\big)\ .$$

Thus it suffices to show $[\phi(s), x_\alpha] \in D\mathfrak{g}$. To do this we use the canonical isomorphism $\mathrm{End}\, V \simeq V \otimes V^* = V_{1,1}$, and apply Consequence 6.6 with $p = q = 1$ and with $A = D\mathfrak{g}$ and $B = \mathfrak{g}$. Making the identification $\mathrm{End}\, V = V_{1,1}$ we have $u_{1,1}(x) = ux - xu = [u,x]$, as remarked before Lemma 6.3, hence $u_{1,1}\mathfrak{g} \subset D\mathfrak{g}$. By 6.6 it follows that $\phi(s)_{1,1}\mathfrak{g} \subset D\mathfrak{g}$, that is, $[\phi(s), x] \in D\mathfrak{g}$ for each $x \in \mathfrak{g}$. q.e.d.

Exercises

1. The class of nilpotent (resp. solvable) Lie algebras is closed under passage to quotient, subalgebras and products. What about extensions?
2. A nilpotent Lie algebra of dimension 2 is abelian. A non-abelian Lie algebra of dimension 2 has a basis $\{X, Y\}$ such that $[X,Y] = X$.
3. A non-abelian nilpotent Lie algebra of dimension 3 has a basis $\{X, Y, Z\}$ such that $[X,Y] = Z$, $[X,Z] = [Y,Z] = 0$.

Chapter VI. Semisimple Lie Algebras

Throughout this chapter, k is a field of characteristic 0, and all algebras and modules are finite dimensional over k.

1. The radical

Let $\mathfrak{g}$ be a Lie algebra. If $\mathfrak{a}$ and $\mathfrak{b}$ are solvable ideals in $\mathfrak{g}$, then $\mathfrak{a}+\mathfrak{b}$ is solvable, because it is an extension of $(\mathfrak{a}+\mathfrak{b})/\mathfrak{a} \approx \mathfrak{b}/(\mathfrak{a}\cap\mathfrak{b})$ by $\mathfrak{a}$. It follows that there exists a solvable ideal in $\mathfrak{g}$ which contains all other solvable ideals. This largest solvable ideal is called the *radical* of $\mathfrak{g}$, and is often denoted by $\mathfrak{r}$.

2. Semisimple Lie algebras

Let $\mathfrak{g}$ be a Lie algebra. One says that $\mathfrak{g}$ is *semisimple* if the radical $\mathfrak{r}$ of $\mathfrak{g}$ is zero. An equivalent condition is that $\mathfrak{g}$ *contains no non-zero abelian ideal*, because if $\mathfrak{r} \neq (0)$, then the last non-zero derived algebra of $\mathfrak{r}$ is a non-zero abelian ideal of $\mathfrak{g}$. Another criterion for semisimplicity is the following:

Theorem 2.1. *$\mathfrak{g}$ is semisimple if and only if its Killing form is non-degenerate.*

Let $\mathfrak{u}$ be the space of all $x \in \mathfrak{g}$ such that $\mathrm{tr}(\mathrm{ad}\, x\, \mathrm{ad}\, y) = 0$ for all $y \in \mathfrak{g}$. It is trivial to check that $\mathfrak{u}$ is an ideal in $\mathfrak{g}$. For $x \in \mathfrak{u}$ we have $\mathrm{tr}(\mathrm{ad}\, x\, \mathrm{ad}\, y) = 0$ for all $y \in \mathfrak{g}$, hence in particular for $y \in D\mathfrak{u}$. By Cartan's criterion, it follows that $\mathrm{ad}_{\mathfrak{g}}\, \mathfrak{u}$ is a solvable Lie subalgebra of $\mathrm{End}(\mathfrak{g})$. Since $\mathrm{ad}_{\mathfrak{g}}\, \mathfrak{u}$ is the quotient of $\mathfrak{u}$ by the center of $\mathfrak{g}$, it follows that $\mathfrak{u}$ itself is solvable. Thus $\mathfrak{u} = 0$ if $\mathfrak{g}$ is semisimple.

To show the converse, we let $\mathfrak{a}$ be an abelian ideal in $\mathfrak{g}$ and will prove that $\mathfrak{a} \subset \mathfrak{u}$. Indeed, let $\sigma = \mathrm{ad}\, x\, \mathrm{ad}\, y$, for $x \in \mathfrak{a}$, $y \in \mathfrak{g}$. Then $\sigma\mathfrak{g} \subset \mathfrak{a}$ and $\sigma\mathfrak{a} = (0)$, hence $\sigma^2 = 0$ and $\mathrm{Tr}\, \sigma = 0$.

Theorem 2.2. *Let $\mathfrak{g}$ be semisimple and let $\mathfrak{a}$ be an ideal in $\mathfrak{g}$. Let $\mathfrak{a}^{\perp}$ be the orthogonal space to $\mathfrak{a}$ with respect to the Killing form of $\mathfrak{g}$. Then $\mathfrak{a}^{\perp}$ is an ideal of $\mathfrak{g}$, and $\mathfrak{g} = \mathfrak{a} \oplus \mathfrak{a}^{\perp}$, direct sum.*

A simple computation, using the invariance of the Killing form, shows that $\mathfrak{a}^{\perp}$ is an ideal. One can show that $\mathfrak{a} \cap \mathfrak{a}^{\perp}$ is solvable, using Cartan's criterion, in the same way we showed $\mathfrak{u}$ solvable in the proof of the preceding theorem. Hence $\mathfrak{a} \cap \mathfrak{a}^{\perp} = (0)$, and the theorem follows.

Definition 2.3. A Lie algebra $\mathfrak{s}$ is called *simple* if

(i) $\mathfrak{s}$ is non-abelian.

(ii) $\mathfrak{s}$ has no ideal other than (0) and $\mathfrak{s}$.

Notice that in the preceding theorem we have $[\mathfrak{a}, \mathfrak{a}^\perp] = 0$, because $\mathfrak{a}$ and $\mathfrak{a}^\perp$ are ideals in $\mathfrak{g}$; hence the decomposition $\mathfrak{g} = \mathfrak{a} \oplus \mathfrak{a}^\perp$ gives an isomorphism of Lie algebras $\mathfrak{g} \simeq \mathfrak{a} \times \mathfrak{a}^\perp$. It follows that any ideal in $\mathfrak{a}$ is an ideal in $\mathfrak{g}$, and consequently $\mathfrak{a}$ is semisimple. Also $\mathfrak{g}/\mathfrak{a} \simeq \mathfrak{a}^\perp$ is semisimple. By induction on $\dim \mathfrak{g}$ one sees therefore:

Corollary 1. *A semisimple Lie algebra is isomorphic to a product of simple Lie algebras.*

If $\mathfrak{s}$ is a simple Lie algebra, then $D\mathfrak{s} = \mathfrak{s}$. Hence:

Corollary 2. *If $\mathfrak{g}$ is semisimple, then $\mathfrak{g} = D\mathfrak{g}$.*

In fact, the decomposition of $\mathfrak{g}$ into a product of simple algebras is *unique*, not only unique up to isomorphism: Let $\mathfrak{g} = \bigoplus \mathfrak{a}_\alpha$, direct sum of simple ideals $\mathfrak{a}_\alpha$, and let $\phi : \mathfrak{g} \to \mathfrak{s}$ be a surjective homomorphism of $\mathfrak{g}$ onto a simple Lie algebra $\mathfrak{s}$. Let ϕ_α be the restriction of ϕ to $\mathfrak{a}_\alpha$. *Claim*: There is an index β such that $\phi_\beta : \mathfrak{a}_\beta \simeq \mathfrak{s}$ is an isomorphism, and $\phi_\alpha = 0$ for $\alpha \neq \beta$. For each α, the image of $\phi_\alpha \mathfrak{a}_\alpha$ is an ideal in $\mathfrak{s}$, because ϕ is surjective and $\mathfrak{a}_\alpha$ an ideal in $\mathfrak{g}$. Hence by the simplicity of $\mathfrak{s}$, ϕ_α is either surjective or zero. If it is surjective, then it is an isomorphism, by the simplicity of $\mathfrak{a}_\alpha$. The set of α's for which ϕ_α is an isomorphism is not empty, because ϕ is surjective. On the other hand that set does not contain two distinct indices $\alpha \neq \beta$, because $[\mathfrak{a}_\alpha, \mathfrak{a}_\beta] = 0$ would imply $[\phi\mathfrak{a}_\alpha, \phi\mathfrak{a}_\beta] = [\mathfrak{s}, \mathfrak{s}] = 0$.

Corollary 3. *If $\mathfrak{g} = \bigoplus \mathfrak{a}_\alpha$ is an expression for $\mathfrak{g}$ as a direct sum of simple ideals $\mathfrak{a}_\alpha$, then any ideal of $\mathfrak{g}$ is a sum of some of the $\mathfrak{a}_\alpha$.*

Examples of semisimple Lie algebras.

1) $\mathfrak{sl}(V)$, the algebra of endomorphisms of V of trace zero is simple if $\dim V \geq 2$.

2) $\mathfrak{sp}(V)$, the algebra of endomorphisms of V leaving invariant a non-degenerate alternating form is simple if $\dim V = 2n$, with $n \geq 1$.

3) $\mathfrak{o}(v)$, the algebra of endomorphisms of V leaving a non-degenerate symmetric form is semisimple for $\dim V \geq 3$, and even simple except if $\dim V = 4$, and the discriminant of the symmetric form is a square.

3. Complete reducibility

Let $\mathfrak{g}$ be a Lie algebra, V a $\mathfrak{g}$-module, and $\varrho : \mathfrak{g} \to \operatorname{End} V$ the corresponding representation.

Definition. V (or ϱ) is called *simple* (or *irreducible*) if $V \neq (0)$ and V has no submodules other than (0) and V.

V (or ϱ) is called *semisimple* (or *completely reducible*) if V is the direct sum of simple submodules, or, what is the same, if every submodule of V has a supplementary submodule.

Danger! $\mathfrak{g}$ may be semisimple as a $\mathfrak{g}$-module without being a semisimple Lie algebra; for example, $\mathfrak{g} = k$.

Theorem (H. Weyl). *If $\mathfrak{g}$ is semisimple, all $\mathfrak{g}$-modules (of finite dimension) are semisimple.*

Remark. Weyl used the "unitarian trick". Let $k = \mathbf{C}$, let G be a connected and simply connected complex Lie group corresponding to $\mathfrak{g}$, and let K be a maximal compact subgroup of G. One proves that any complex group submanifold of G containing K is equal to G. Hence G-submodules of V are the same as K-submodules; since K is compact, there exists a K-invariant definite Hermitian form on V, with which to construct orthogonal supplementary subspaces. In case $G = \mathrm{SL}(n)$, one can take $K = \mathrm{SU}(n)$, the special unitary group; hence the name "unitarian trick". A purely algebraic proof of Weyl's theorem was found only several years later.

We now prove the theorem in a sequence of steps.

Step 1. If $\mathfrak{g}$ is semisimple, and $\varrho : \mathfrak{g} \to \operatorname{End} V$ is injective, then the form $B_\varrho(x, y) = \operatorname{Tr}_V(\varrho(x)\varrho(y))$ is non-degenerate. Indeed, by Cartan's criterion, the ideal of all $x \in \mathfrak{g}$ such that $B_\varrho(x, y) = 0$ for all $y \in \mathfrak{g}$ is solvable, hence 0.

Step 2. Let B be a non-degenerate invariant symmetric bilinear form on a Lie algebra $\mathfrak{g}$. Let (e_i) and (f_j) be bases for $\mathfrak{g}$ which are dual with respect to B, that is, such that $B(e_i, f_j) = \delta_{ij}$ (Kronecker delta). Let $b = \sum e_i f_j$ in $U\mathfrak{g}$. Then b is in the center of $U\mathfrak{g}$, and is independent of the choice of e_i, f_j. Indeed, the map $\mathfrak{g} \otimes_k \mathfrak{g} \to \operatorname{Hom}_k(\mathfrak{g}, \mathfrak{g})$ for which $x \otimes y \mapsto (z \mapsto B(y, z)x)$ is an isomorphism because B is non-degenerate, and is a homomorphism of $\mathfrak{g}$-modules as one readily checks. Moreover, it carries $\sum e_i \otimes f_j$ onto the identity homomorphism 1. Thus, the element b is the image of 1 under the composition of the $\mathfrak{g}$-homomorphisms

$$\operatorname{Hom}(\mathfrak{g}, \mathfrak{g}) \simeq \mathfrak{g} \otimes \mathfrak{g}^* \xrightarrow{B} \mathfrak{g} \otimes \mathfrak{g} \longrightarrow U\mathfrak{g}\,.$$

Since 1 is killed by $\mathfrak{g}$, b is also, and since $\mathfrak{g}$ generates $U\mathfrak{g}$, it follows that b is in the center of $U\mathfrak{g}$ as contended. This element b is called the *Casimir element* corresponding to B.

Step 3. The situation being as in Step 1, let b be the Casimir element corresponding to B_ϱ. Then b defines an endomorphism of the $\mathfrak{g}$-module V, and we have $\operatorname{Tr}_V(b) = \dim \mathfrak{g}$. Indeed, b commutes with the action of $\mathfrak{g}$ on V because it is in the center of $U\mathfrak{g}$. To compute its trace we have $\operatorname{Tr}(b) = \sum \operatorname{Tr}(\varrho(e_i)\varrho(f_i)) = \sum B(e_i, f_i) = \dim \mathfrak{g}$.

Step 4. If the $\mathfrak{g}$-module V in Step 3 is simple, then $\varrho(b)$ is an automorphism of V, *unless* $\mathfrak{g} = 0$ (in which case V is one-dimensional). Indeed, by "Schur's lemma", an endomorphism of a simple module is either an automorphism, or zero, and $\varrho(b)$ is not zero unless $\mathfrak{g} = 0$ because $\operatorname{Tr} \varrho(b) = \dim \mathfrak{g}$ and k is of characteristic zero.

Step 5. Let $0 \to V \to W \to k \to 0$ be an exact sequence of $\mathfrak{g}$-modules, with $\mathfrak{g}$ acting trivially on k (in fact, there is no other possibility since $\mathfrak{g} = D\mathfrak{g}$; we are supposing $\mathfrak{g}$ semisimple). We shall prove that the sequence splits, that is, that there exists a line in W stable by $\mathfrak{g}$ and supplementary to V, i.e., mapping onto k. This special case of the theorem, the so-called "lifting of invariants" principle, is the critical case, to which the general case can be reduced by use of modules of homomorphisms (see below). We break this step 5 into three substeps.

Step 5a. Reduction to case V is a simple $\mathfrak{g}$-module. This is easily accomplished by induction on $\dim V$. If $V_1 \subset V$, with $0 \neq V_1$, and $V_1 \neq V$, then by considering the sequence $0 \to V/V_1 \to W/V_1 \to k$ we would obtain a supplementary line $\tilde{V}/V_1$ to V/V_1 in W/V_1, and then from the sequence $0 \to V_1 \to \tilde{V} \to k \to 0$ we would obtain a supplementary line to V_1 in $\tilde{V}$ which, by construction, would be a supplementary line to V in W.

Step 5b. Reduction to case $\mathfrak{g}$ operates faithfully on V. Let $\mathfrak{a} = \mathrm{Ker}(\mathfrak{g} \to \mathrm{End}\, V)$. For $x \in \mathfrak{a}$ we have $xW \subset V$ and $xV = (0)$. Hence $D\mathfrak{a}$ kills W. But $D\mathfrak{a} = \mathfrak{a}$, because an ideal in a semisimple $\mathfrak{g}$ is semisimple. Hence $\mathfrak{g}/\mathfrak{a}$ acts on W, and by construction it acts faithfully on V. Of course, we have not lost the semisimplicity of $\mathfrak{g}$, because a quotient of a semisimple algebra is semisimple.

Step 5c. Assume V simple and $\varrho : \mathfrak{g} \to \mathrm{End}\, V$ is injective. The associated bilinear form B_ϱ is non-degenerate; let $b \in U\mathfrak{g}$ be the corresponding Casimir element. It furnishes a $\mathfrak{g}$-endomorphism of W, and $bW \subset V$ because b kills k. If $\mathfrak{g} = (0)$ there is no problem. Otherwise, by step 4, we have $bV = V$, and it follows that $\mathrm{Ker}(b : W \to W)$ is a supplementary line to V in W stable by $\mathfrak{g}$.

Step 6. The general case. Let $0 \to E_1 \to E \to E_2 \to 0$ be an exact sequence of $\mathfrak{g}$-modules. We must show that it splits. Let W be the subspace of $\mathrm{Hom}_k(E, E_1)$ consisting of the elements whose restriction to E_1 is a homothety, and let V be the subspace whose restriction to E_1 is zero. There results an exact sequence $0 \to V \to W \to k \to 0$ (unless $E_1 = (0)$, in which case there is no problem anyway). Applying step 5, we get an element $\phi \in W$ which is invariant by $\mathfrak{g}$ and maps onto 1 in k, that is a $\mathfrak{g}$-homomorphism $E \to E_1$ whose restriction to E_1 is 1. q.e.d.

From the point of view of homological algebra, step 5 amounts to proving $\mathrm{Ext}^1_U(k, V) = 0$, where $U = U\mathfrak{g}$, and this is accomplished in step 5c by computing the action of the central element b on Ext^1 in two ways. Since b kills k, it kills Ext^1, and since b is an automorphism of V, it gives an automorphism of Ext^1. Hence $\mathrm{Ext}^1 = 0$. In general, one defines $H^r(\mathfrak{g}, V) = \mathrm{Ext}^r_U(k, V)$. Step 6 amounts to showing $\mathrm{Ext}^1_U(E_2, E_1) = H^1(\mathfrak{g}, \mathrm{Hom}_k(E_2, E_1)) = 0$.

Corollary 1. *Let $\mathfrak{g}$ be a semisimple ideal of a Lie algebra $\mathfrak{h}$. Then there exists a unique ideal $\mathfrak{a}$ in $\mathfrak{h}$ such that $\mathfrak{h} = \mathfrak{g} \oplus \mathfrak{a}$ (direct sum).*

Applying complete reducibility to $\mathfrak{h}$ as a $\mathfrak{g}$-module we get a k-subspace $\mathfrak{a}$ of $\mathfrak{h}$ supplementary to $\mathfrak{g}$ and stable by $\mathrm{ad}\, x$ for $x \in \mathfrak{g}$. I claim $[\mathfrak{g}, \mathfrak{a}] = 0$;

indeed, $[\mathfrak{g},\mathfrak{a}] \subset \mathfrak{g}$ because $\mathfrak{g}$ is an ideal, and $[\mathfrak{g},\mathfrak{a}] \subset \mathfrak{a}$ because $\mathfrak{a}$ is stable by $\mathfrak{g}$. It follows that $\mathfrak{a}$ consists *exactly* of those $y \in \mathfrak{h}$ such that $[\mathfrak{g}, y] = (0)$, because, writing $y = x + a$, with $x \in \mathfrak{g}$, $a \in \mathfrak{a}$, we have $[\mathfrak{g}, y] = [\mathfrak{g}, x]$, and $[\mathfrak{g}, x] = (0)$ implies $x = 0$ because the center of $\mathfrak{g}$ is zero. This shows that $\mathfrak{a}$ is unique, even as $\mathfrak{g}$-submodule, and also that $\mathfrak{a}$ is an ideal in $\mathfrak{h}$, because it is the annihilator of the $\mathfrak{h}$-module $\mathfrak{g}$.

Corollary 2. *If $\mathfrak{g}$ is semisimple, then every derivation of $\mathfrak{g}$ is of the form* $\operatorname{ad} x$, *with* $x \in \mathfrak{g}$.

Apply the preceding corollary with $\mathfrak{h} = \operatorname{Der}(\mathfrak{g})$, the Lie algebra of derivations of $\mathfrak{g}$. It is true that $\mathfrak{g}$ is an ideal in $\operatorname{Der}(\mathfrak{g})$, because for $x \in \mathfrak{g}$ and $D \in \operatorname{Der}(\mathfrak{g})$, we have $[D, \operatorname{ad} x] = \operatorname{ad}(Dx)$. Hence $\operatorname{Der}(\mathfrak{g}) = \mathfrak{g} \oplus \mathfrak{a}$, where $\mathfrak{a}$ consists of the derivations commuting with $\operatorname{ad} \mathfrak{g}$. Let $D \in \mathfrak{a}$. Then $\operatorname{ad}(Dx) = [D, \operatorname{ad} x] = 0$. Hence $Dx = 0$, because the center of $\mathfrak{g}$ is zero. Hence $\mathfrak{a} = 0$. q.e.d.

4. Levi's theorem

Let $\mathfrak{g}$ be a Lie algebra.

Theorem 4.1 (Levi). *Let $\phi : \mathfrak{g} \to \mathfrak{s}$ be a surjective homomorphism of $\mathfrak{g}$ onto a semisimple Lie algebra $\mathfrak{s}$. Then there exists a homomorphism $\varepsilon : \mathfrak{s} \to \mathfrak{g}$ such that $\phi \circ \varepsilon = 1_{\mathfrak{s}}$.*

Let $\mathfrak{a} = \operatorname{Ker} \phi$, and write $\mathfrak{s} = \mathfrak{g}/\mathfrak{a}$. The *crucial case* of the theorem is that in which $\mathfrak{a}$ is abelian, and is a simple $\mathfrak{g}$- (or $\mathfrak{s}$-) module with non-trivial action. The first step of the proof is the *reduction to the crucial case*. Suppose $\mathfrak{a}_1$ is an ideal in $\mathfrak{g}$, and $0 \subset \mathfrak{a}_1 \subset \mathfrak{a}$. If we can find a supplementary subalgebra $\mathfrak{s}_1 = \mathfrak{g}_1/\mathfrak{a}_1$ to $\mathfrak{a}/\mathfrak{a}_1$ in $\mathfrak{g}/\mathfrak{a}_1$, and a supplementary subalgebra $\mathfrak{s}_2$ to $\mathfrak{a}_1$ in $\mathfrak{g}_1$, then $\mathfrak{s}_2$ is supplementary to $\mathfrak{a}$ in $\mathfrak{g}$. Hence, by induction on $\dim \mathfrak{a}$, we may suppose $\mathfrak{a}$ is a simple $\mathfrak{g}$-module. The radical $\mathfrak{r}$ of $\mathfrak{g}$ is in $\mathfrak{a}$. If $\mathfrak{r} = 0$, then $\mathfrak{g}$ is semisimple, and we are done, by Theorem 2.2. If $\mathfrak{r} = \mathfrak{a}$, then $\mathfrak{a}$ is solvable, hence $\mathfrak{a} \neq [\mathfrak{a},\mathfrak{a}]$. But $[\mathfrak{a},\mathfrak{a}]$ is an ideal in $\mathfrak{g}$, so $[\mathfrak{a},\mathfrak{a}] = 0$, i.e., $\mathfrak{a}$ is abelian. If $\mathfrak{g}$ acts trivially on $\mathfrak{a}$, then $\mathfrak{a}$ is in the center of $\mathfrak{g}$, hence $\mathfrak{g}$ operates on $\mathfrak{g}$ through $\mathfrak{g}/\mathfrak{a} \simeq \mathfrak{s}$, and $\mathfrak{g}$ is completely reducible as an $\mathfrak{s}$-module, so there is an ideal supplementary to $\mathfrak{a}$.

Assume now we are in the *crucial case*: $\mathfrak{a}$ abelian, and a simple $\mathfrak{s}$-module with non-trivial action. If we had cohomology at our disposal, and knew that the extensions of $\mathfrak{s}$ by $\mathfrak{a}$ are classified by $H^2(\mathfrak{s},\mathfrak{a}) = \operatorname{Ext}^2_{U\mathfrak{s}}(k,\mathfrak{a})$, we would be finished, because we could use a Casimir element to show that the Ext group is zero. But not having cohomology, we resort to the following argument of Bourbaki:

Lemma. *Let W be a $\mathfrak{g}$-module. Suppose an element $w \in W$ satisfies the conditions*

a) *the map* $a \mapsto aw$ *is a bijection* $\mathfrak{a} \xrightarrow{\sim} \mathfrak{a}w$;

b) $\mathfrak{g}w = \mathfrak{a}w$.

Let $\mathfrak{i}_w = \{ x \in \mathfrak{g} \mid xw = 0 \}$ *be the stabilizer of* w. *Then* $\mathfrak{i}_w$ *is a Lie subalgebra of* $\mathfrak{g}$, *and* $\mathfrak{g} = \mathfrak{a} \oplus \mathfrak{i}_w$ (direct sum as vector spaces).

The lemma is completely trivial. Our problem now is to construct a suitable w. We let $W = \operatorname{End}(\mathfrak{g})$, viewed as $\mathfrak{g}$-module in the usual way, the representation $\sigma : \mathfrak{g} \to \operatorname{End} W = \operatorname{End}\operatorname{End}\mathfrak{g}$ being defined by $\sigma(x)\phi = \operatorname{ad} x \circ \phi - \phi \circ \operatorname{ad} x = [\operatorname{ad} x, \phi]$. We define three subspaces $P \subset Q \subset R \subset W$ as follows:

$$P = \{ \operatorname{ad}_{\mathfrak{g}} a \mid a \in \mathfrak{a} \}$$
$$Q = \{ \phi \in W \mid \phi\mathfrak{g} \subset \mathfrak{a} \text{ and } \phi\mathfrak{a} = 0 \}$$
$$R = \{ \phi \in W \mid \phi\mathfrak{g} \subset \mathfrak{a} \text{ and } \phi \mid \mathfrak{a} \text{ is a homothety} \}.$$

We leave to the reader the task of showing that these are $\mathfrak{g}$-submodules of W. We have an exact sequence of $\mathfrak{g}$-modules

$$0 \longrightarrow Q \xrightarrow{i} R \xrightarrow{\varrho} k \to 0$$

where i is the inclusion, and ϱ the map which associates with each $r \in R$ the scalar by which r multiplies elements of $\mathfrak{a}$. If $x \in \mathfrak{a}$ and $\phi \in R$, then $\sigma(x)\phi = \operatorname{ad} x \circ \phi - \phi \circ \operatorname{ad} x = -\lambda \operatorname{ad} x$, where $\lambda = \varrho(\phi) \in k$. Thus, $\sigma(x)R \subset P$, for $x \in \mathfrak{a}$, and the exact sequence

$$0 \longrightarrow Q/P \longrightarrow R/P \xrightarrow{\bar{\varrho}} k \to 0$$

may be viewed as a sequence of $\mathfrak{s}$-modules. By the principle of lifting invariants, there exists $\bar{w} \in R/P$ such that $\bar{\varrho}(\bar{w}) = 1$, and such that $\bar{w}$ is invariant by $\mathfrak{s}$. Let w be an inverse of image of $\bar{w}$ in R. We contend that w satisfies the conditions of the lemma above.

a) Let $a \in \mathfrak{a}$. Then $\sigma(a)w = -\operatorname{ad} a$. If $\sigma(a)w = 0$, then $\operatorname{ad}_{\mathfrak{g}} a = 0$, that is, $[a, x] = 0$ for all $x \in \mathfrak{g}$. This implies $a = 0$, because $\mathfrak{a}$ is simple, and $\mathfrak{g}$ acts non-trivially.

b) Let $x \in \mathfrak{g}$. We must show that $\sigma(x)w$ is of the form $\sigma(a)w$ for some $a \in \mathfrak{a}$. Since $\sigma(a)w = -\operatorname{ad}_{\mathfrak{g}} a$, this amount to showing $\sigma(x)w \in P$. But that is just the invariance of $\bar{w}$. q.e.d.

Corollary 1. *An arbitrary Lie algebra* $\mathfrak{g}$ *is the semi-direct product of its radical* $\mathfrak{r}$ *and a semisimple subalgebra.*

One applies Theorem 4.1 to $\mathfrak{g} \to \mathfrak{g}/\mathfrak{r}$.

Remark. This corollary has a complement, due to Malcev, which says that, if $\mathfrak{s}_1$ and $\mathfrak{s}_2$ are two subalgebras of $\mathfrak{g}$ such that $\mathfrak{r} \oplus \mathfrak{s}_i = \mathfrak{g}$, there exists an automorphism σ of $\mathfrak{g}$ such that $\sigma(\mathfrak{s}_1) = \mathfrak{s}_2$ [one can even choose σ of the special form $e^{\operatorname{ad}(a)}$, where $a \in \mathfrak{r}$, and $\operatorname{ad}(a)$ is nilpotent]. When $\mathfrak{r}$ is abelian, this amounts to the vanishing of $H^1(\mathfrak{g}/\mathfrak{r}, \mathfrak{r})$; the general case follows by "dévissage". See Bourbaki for more details.

If $\mathfrak{g}$ is a Lie algebra such that $\mathfrak{g} \neq D\mathfrak{g}$, and if $\mathfrak{a}$ is a subspace of $\mathfrak{g}$ of codimension 1 containing $D\mathfrak{g}$, then $\mathfrak{a}$ is an ideal in $\mathfrak{g}$, and we have $\mathfrak{g} = \mathfrak{a} \oplus kx$ for any $x \notin \mathfrak{a}$. Since kx is automatically a Lie subalgebra we get:

Corollary 2. *A non-zero Lie algebra which is neither simple nor one-dimensional abelian is a semidirect product of two Lie algebras of smaller dimensions.*

5. Complete reducibility continued

The following theorem gives a criterion for the complete reducibility of a representation of a Lie algebra.

Theorem 5.1. *Suppose k is algebraically closed. Let V be a vector space and let $\mathfrak{g}$ be a Lie subalgebra of* End V. *Then V is completely reducible as a $\mathfrak{g}$-module if and only if the following two conditions are satisfied.*

a) *$\mathfrak{g}$ is a product $\mathfrak{c} \times \mathfrak{s}$ where $\mathfrak{c}$ is abelian and $\mathfrak{s}$ semisimple.*

b) *the elements of $\mathfrak{c}$ can be put in diagonal form by a suitable choice of basis for V.*

Remarks. 1) If k is not algebraically closed, the same statement holds if we replace the condition b) by the statement that elements of $\mathfrak{c}$ are *semisimple* (i.e., diagonalizable over the algebraic closure!).

2) The ambiguity in statement b) is only apparent. If each element of $\mathfrak{c}$ is individually diagonalizable, then they are all simultaneously diagonalizable, because $\mathfrak{c}$ is commutative.

Suppose V is completely reducible as a $\mathfrak{g}$-module. Let $\mathfrak{r}$ be the radical of $\mathfrak{g}$. By Lie's theorem (Chap. V.5), there exists a line in V stable by $\mathfrak{r}$ (unless $V = (0)$, in which case there is nothing to prove), or, what is the same, there exists a linear form $\chi : \mathfrak{r} \to k$ such that its eigenspace

$$V_\chi = \{ v \in V \mid xv = \chi(x)v \text{ for all } x \in \mathfrak{r} \}$$

is non-zero. By the "main lemma" used in the proof of Lie's theorem (*loc. cit.*), V_χ is stable under $\mathfrak{g}$. By complete reducibility, we conclude that there exist characters χ_i of $\mathfrak{r}$ such that

$$(*) \qquad V = V_{\chi_1} \oplus V_{\chi_2} \oplus \cdots \oplus V_{\chi_m} \qquad \text{(direct sum)}.$$

From this decomposition it is clear first that $\mathfrak{r}$ acts diagonally, and commutes with the action of $\mathfrak{g}$. Thus, $\mathfrak{r} = \mathfrak{c}$ is the center of $\mathfrak{g}$. To get $\mathfrak{s}$, we can either quote Levi's theorem, or argue directly using the adjoint representation.

Conversely, suppose conditions a) and b) are satisfied. By b) we have a decomposition of the vector space V of the form $(*)$, where the χ_i's are characters (linear forms) on $\mathfrak{c}$. Since $\mathfrak{c}$ is in the center of $\mathfrak{g} = \mathfrak{c} \times \mathfrak{s}$, the eigenspaces V_{χ_i} are stable under $\mathfrak{g}$. We are therefore reduced to the case $V = V_\chi$. But then $\mathfrak{c} \times \mathfrak{s}$-submodules are the same as $\mathfrak{s}$-submodules, and we are done by Weyl's Theorem.

Corollary 1. *Suppose $\mathfrak{g} = \mathfrak{c} \times \mathfrak{s}$ with $\mathfrak{c}$ abelian and $\mathfrak{s}$ semisimple. A $\mathfrak{g}$-module W is semisimple if and only if $\mathfrak{c}$ acts diagonally on W.*

Corollary 2. *Let $\mathfrak{g}$ be an arbitrary Lie algebra, and V a $\mathfrak{g}$-module. If V is completely reducible, then so are the tensor modules*

$$V_{p,q} = \bigotimes^p V \bigotimes^q V^* .$$

Indeed, the image $\bar{\mathfrak{g}}$ of $\mathfrak{g}$ in $\operatorname{End} V$ is of the form $\mathfrak{c} \times \mathfrak{s}$, with $\mathfrak{c}$ acting diagonally on V, and hence on all $V_{p,q}$.

Corollary 3. *The tensor product of completely reducible $\mathfrak{g}$-modules is completely reducible.*

Using these results we can prove:

Theorem 5.2. *Let V be a finite dimensional vector space over k. Let $\mathfrak{g} \subset \operatorname{End} V$ be a Lie algebra of endomorphisms of V. If $\mathfrak{g}$ is semisimple, then $\mathfrak{g}$ is determined by its tensor invariants, that is, there exist some elements $v_\alpha \in V_{p,q}$* (for various (p,q)'s; we should write (p_α, q_α)) *such that $\mathfrak{g} = \{ x \in \operatorname{End} V \mid xv_\alpha = 0 \text{ for all } \alpha \}$.*

By a standard linear argument, we can reduce the question to the case where k is algebraically closed. Let $\mathfrak{h}$ be the set of all $x \in \operatorname{End} V$ such that $xv = 0$ for every v in some $V_{p,q}$ such that $\mathfrak{g}v = 0$. Clearly $\mathfrak{g} \subset \mathfrak{h} \subset \operatorname{End} V$, and $\mathfrak{h}$ is a Lie algebra. Our task is to show $\mathfrak{h} = \mathfrak{g}$.

Step 1: If a linear map $u : V_{p,q} \to V_{r,s}$ is a $\mathfrak{g}$-homomorphism, then it is an $\mathfrak{h}$-homomorphism, because we can identify $\operatorname{Hom}_k(V_{p,q}, V_{r,s})$ with $V_{q+r,p+s}$, as $\operatorname{End} V$ modules, and for a linear map to be a $\mathfrak{g}$-homomorphism is the same as for it to be killed by the action of $\mathfrak{g}$.

Step 2: If a subspace $W \subset V_{p,q}$ is stable under $\mathfrak{g}$, it is stable under $\mathfrak{h}$. Indeed, since $V_{p,q}$ is completely reducible as a $\mathfrak{g}$-module, there is a $\mathfrak{g}$-endomorphism u of $V_{p,q}$ projecting $V_{p,q}$ into W. Since u is also an $\mathfrak{h}$-endomorphism, its image W is stable under $\mathfrak{h}$.

Step 3: We have $\mathfrak{h} = \mathfrak{g} \times \mathfrak{c}$, where $\mathfrak{c}$ is the center of $\mathfrak{h}$. For, by Step 2, with $W = \mathfrak{g}$, and $p = q = 1$, we see that $\mathfrak{g}$ is an ideal in $\mathfrak{h}$. By Corollary 1 of Weyl's Theorem (VI.3) we have $\mathfrak{h} = \mathfrak{g} \times \mathfrak{c}$, where $\mathfrak{c}$ is an ideal in $\mathfrak{h}$ commuting with

$\mathfrak{g}$. By Step 1, it follows that $\mathfrak{c}$ commutes with $\mathfrak{h}$, that is, $\mathfrak{c}$ is in the center of $\mathfrak{h}$.

Step 4: Let W be an irreducible $\mathfrak{g}$-submodule of V. Then W is stable under $\mathfrak{c}$, and by "Schur's Lemma", the elements of $\mathfrak{c}$ act as homotheties on W. We must show that these homotheties are zero; since V is the direct sum of W's, this will show $\mathfrak{c} = 0$ and conclude our proof. Since we are in characteristic zero, we can show that a homothety is zero by showing that its trace is zero.

Lemma 5.3. *Let $\mathfrak{g}$ be a Lie algebra and W a $\mathfrak{g}$-module of dimension m. Then its m-th exterior power $\bigwedge^m W$, as a quotient space of $\otimes^m W$ (or as a subspace in characteristic zero) is stable under $\mathfrak{g}$, and an element $x \in \mathfrak{g}$ acts on the one-dimensional space $\bigwedge^m W$ by the scalar $\mathrm{Tr}_W(x)$.*

We leave the proof of this Lemma as an exercise. Granting the lemma, we argue as follows. We have $\bigwedge^m W \subset \otimes^m W \subset \otimes^m V = v_{m,0}$. Since our semisimple $\mathfrak{g}$ has no non-trivial one dimensional module ($D\mathfrak{g} = \mathfrak{g}$), we conclude that $\bigwedge^m W$ is killed by $\mathfrak{g}$. Hence it is killed by $\mathfrak{c}$, and hence $\mathrm{Tr}_W(x) = 0$ for all $x \in \mathfrak{g}$. This concludes the proof.

Corollary 5.4. *Let $\mathfrak{g} \subset \mathrm{End}(V)$ be semisimple. Let $x \in \mathfrak{g}$ and let $x = n+s$ be the canonical decomposition of x, with n nilpotent, s semisimple and $[n, s] = 0$* (cf. Chap. V). *Then:*

a) *n and s belongs to $\mathfrak{g}$.*

b) *For any $\phi \in \mathrm{Hom}_{\mathbf{Q}}(k, k)$, $\phi(s)$ belongs to $\mathfrak{g}$.*

This follows from the fact that any element in $V_{p,q}$ killed by $\mathfrak{g}$ is killed by x and hence also by n, s and $\phi(s)$.

Definition 5.5. Let $\mathfrak{g}$ be a semisimple Lie algebra. An element $x \in \mathfrak{g}$ is called *semisimple* (resp. *nilpotent*) if $\mathrm{ad}(x)$ is *semisimple* (resp. *nilpotent*).

Theorem 5.6. *If $\mathfrak{g}$ is semisimple, any $x \in \mathfrak{g}$ can be uniquely written $x = n + s$, with $n \in \mathfrak{g}$, $s \in \mathfrak{g}$, n nilpotent, s semisimple and $[n, s] = 0$.*

This follows from Corollary 5.4, applied to the adjoint representation (i.e., $V = \mathfrak{g}$).

Theorem 5.7. *If $\phi : \mathfrak{g}_1 \to \mathfrak{g}_2$ is a homomorphism of semisimple Lie algebras, and if $x \in \mathfrak{g}_1$ is semisimple (resp. nilpotent), so is $\phi(x)$.*

Notice first that $\mathfrak{g}_2$ can be made into a $\mathfrak{g}_1$-module *via* ϕ. Let V be the product of the $\mathfrak{g}_1$-modules $\mathfrak{g}_1$ and $\mathfrak{g}_2$. Applying Corollary 5.4 to V, we see that any $x \in \mathfrak{g}_1$ can be written $x = n + s$, where $n \in \mathfrak{g}_1$, $s \in \mathfrak{g}_1$, $[n, s] = 0$, $\mathrm{ad}(n)$ and $\mathrm{ad}(\phi(n))$ nilpotent, $\mathrm{ad}(s)$ and $\mathrm{ad}(\phi(s))$ semisimple. If x is semisimple (resp. nilpotent), this implies $n = 0$ (resp. $s = 0$), hence $\phi(x)$ is semisimple (resp. nilpotent).

6. Connection with compact Lie groups over R and C

We begin with:

Theorem 6.1. *Let G be a connected compact Lie group over* $\mathbf{C}$. *Then $\mathfrak{g}$ is a complex torus, that is, $\mathfrak{g}$ is of the form $\mathbf{C}^n/\Gamma$, where Γ is a discrete subgroup of rank $2n$ in $\mathbf{C}^n$.*

By the maximum principle, there is no non-constant analytic function on G, and hence no non-constant analytic map of G into $\mathrm{End}_{\mathbf{C}}\,\mathfrak{g} \simeq \mathbf{C}^{n^2}$ where $n = \dim \mathfrak{g}$ and $\mathfrak{g}$ is the Lie algebra of G. The inner automorphism $x \mapsto gxg^{-1}$ induced by an element $g \in G$ induces an automorphism of $\mathfrak{g}$ which is denoted by $\mathrm{Ad}\, g$. The map $g \mapsto \mathrm{Ad}\, g \in \mathbf{C}^{n^2}$ is analytic, hence constant, so $\mathrm{Ad}\, g = \mathrm{Ad}\, 1 = 1$ for all $g \in G$. For x near zero in $\mathfrak{g}$ we have $g(\exp x)g^{-1} = \exp(\mathrm{Ad}\, g(x))$, and since the exponential mapping is a homeomorphism of a neighborhood of zero in $\mathfrak{g}$ onto a neighborhood of 1 in G, we conclude that G is locally abelian. Since G is connected, G is abelian. Hence the universal covering of G is $\mathbf{C}^n$, and $G \simeq \mathbf{C}^n/\Gamma$, with Γ discrete, as contended. Since G is compact, Γ is a lattice of maximal rank $2n$.

Theorem 6.2. *Let G be a compact Lie group over* $\mathbf{R}$ *with Lie algebra $\mathfrak{g}$. Then $\mathfrak{g} \simeq \mathfrak{c} \times \mathfrak{s}$, where $\mathfrak{c}$ is abelian, and $\mathfrak{s}$ semisimple with negative definite Killing form.*

We shall also prove a converse:

Theorem 6.3. *If $\mathfrak{g}$ is a Lie algebra over* $\mathbf{R}$ *such that $\mathfrak{g} \simeq \mathfrak{c} \times \mathfrak{s}$ with $\mathfrak{c}$ abelian and $\mathfrak{s}$ semisimple with a definite Killing form then there exists a compact Lie group over* $\mathbf{R}$ *giving $\mathfrak{g}$. Moreover, if $\mathfrak{c} = 0$, then any connected G giving $\mathfrak{g}$ is compact.*

Proof of Theorem 2. As explained in the proof of Theorem 6.1, G acts on $\mathfrak{g}$ by Ad, and, since G is compact, there exists a Euclidean structure (positive definite quadratic form) on $\mathfrak{g}$ which is left fixed by G, and hence by $\mathfrak{g}$. Hence, $\mathfrak{g}$ is completely reducible as a $\mathfrak{g}$-module. It follows that $\mathfrak{g}$ is the direct sum of minimal non-zero ideals $\mathfrak{a}_i$ and is therefore isomorphic to the product of the $\mathfrak{a}_i$. Each $\mathfrak{a}_i$ is either simple or one-dimensional abelian. Hence $\mathfrak{g} \simeq \mathfrak{c} \times \mathfrak{s}$ with $\mathfrak{c}$ abelian and $\mathfrak{s}$ semisimple. It remains to show that the Killing form of $\mathfrak{s}$ is negative definite. Let (x, y) denote the Euclidean inner product on $\mathfrak{g}$. For $x \in \mathfrak{s}$ let $u = \mathrm{ad}_{\mathfrak{s}}\, x$. For $y, z \in \mathfrak{s}$ we have then $(uy, z) + (y, uz) = 0$, because the Euclidean structure on $\mathfrak{s}$ is invariant. Putting $z = uy$, we find $(y, u^2 y) = -(uy, uy)$. Let (y_i) be an orthonormal basis for $\mathfrak{s}$. We have:

$$\mathrm{Tr}_{\mathfrak{s}}(u^2) = \sum_i (y_i, u^2 y_i) = -\sum_i |uy_i|^2 .$$

If $x \neq 0$, then $u = \mathrm{ad}(x) \neq 0$ (because the center of $\mathfrak{s}$ is zero), hence $\mathrm{Tr}_{\mathfrak{s}}(u^2) < 0$. This proves that the Killing form of $\mathfrak{s}$ is negative definite.

Let us now prove Theorem 6.3. As a compact Lie group over $\mathbf{R}$ giving $\mathfrak{c}$ we can take a real torus, $(\mathbf{R}/\mathbf{Z})^n$. To get one giving $\mathfrak{s}$, we take $\mathrm{Aut}\,\mathfrak{s}$, which is a closed subgroup of the orthogonal group of linear transformations of $\mathfrak{s}$ leaving fixed the Killing form of $\mathfrak{s}$. Since that form is definite, the latter group, and hence $\mathrm{Aut}\,\mathfrak{s}$, is compact. The Lie algebra of $\mathrm{Aut}\,\mathfrak{s}$ is the algebra of derivations of $\mathfrak{s}$, which is isomorphic to $\mathfrak{s}$ by Corollary 2 of Weyl's Theorem (VI.3). This proves the first part of Theorem 6.3.

Suppose now $\mathfrak{c} = 0$ (hence $\mathfrak{g}$ is semisimple) and let G be a connected Lie group with Lie algebra $\mathfrak{g}$. We have a canonical homomorphism:

$$\mathrm{Ad} : G \to \mathrm{Aut}\,\mathfrak{g} .$$

We have just seen that $\mathrm{Aut}\,\mathfrak{g}$ is a compact Lie group with Lie algebra $\mathfrak{g}$; hence Ad is étale. Let $H = \mathrm{Im}(\mathrm{Ad}) =$ connected component of $\mathrm{Aut}\,\mathfrak{g}$, and let $Z = \mathrm{Ker}(\mathrm{Ad})$. We have $G/Z = H$, Z is discrete, H is compact, and the commutator group (H, H) is dense in H (this follows *via* Lie theory, from the fact that $\mathfrak{g} = [\mathfrak{g}, \mathfrak{g}]$). Hence G is compact (cf. Bourbaki, *Int.*, VII, §3, Prop. 5).

Exercises

1. Let $\mathfrak{g}$ be a Lie algebra, let $\mathfrak{r}$ be its radical, and let $\mathfrak{i}$ be the intersection of the kernels of the irreducible representation of $\mathfrak{g}$.

   a) Show that $\mathfrak{i} = [\mathfrak{g}, \mathfrak{r}] = D\mathfrak{g} \cap \mathfrak{r}$. (Hint: use Levi's theorem to prove that $[\mathfrak{g}, \mathfrak{r}] = D\mathfrak{g} \cap \mathfrak{r}$.)

   b) Show that $x \in \mathfrak{r}$ belongs to $\mathfrak{i}$ if and only if $\varrho(x)$ is nilpotent for every representation ϱ of $\mathfrak{g}$.

2. Let $\mathfrak{g}$ be a Lie algebra and let $B(x, y)$ be a non-degenerate invariant symmetric bilinear form on $\mathfrak{g}$.

   a) Let $x, y \in \mathfrak{g}$. Show the equivalence of:

   (i) $y \in \mathrm{Im}\,\mathrm{ad}(x)$.

   (ii) $B(y, z) = 0$ for all z which commute with x.

   b) Assume $\mathfrak{g}$ semisimple. Let $x \in \mathfrak{g}$ be such that $\mathrm{ad}(x)$ is nilpotent. Show that there exists $h \in \mathfrak{g}$ such that $[h, x] = x$. Use this to prove that $\varrho(x)$ is nilpotent for any representation ϱ of $\mathfrak{g}$.

3. Give an example of a Lie algebra $\mathfrak{g}$, with a non-zero radical, and a non-degenerate invariant symmetric bilinear form.

4. Let $\mathfrak{g}$ be a Lie algebra and let V be an irreducible $\mathfrak{g}$-module. Let K be the ring of $\mathfrak{g}$-endomorphisms of V. Show that K is a field. Give an example where K is not commutative.

5. Let $\mathfrak{g}$ be a semisimple Lie algebra, and let K be the ring of $\mathfrak{g}$-endomorphisms of $\mathfrak{g}$ (with the adjoint representation). Let $\bar{k}$ be an algebraic closure of k.

   a) Assume $k = \bar{k}$ and $\mathfrak{g} = \bigoplus_{i=1}^{i=h} \mathfrak{s}_i$, where $\mathfrak{s}_i$ are simple. Show that K is isomorphic to the product of h copies of k.

   b) No assumption on k (except $\operatorname{char} k = 0$, of course). Show that $[K : k] = h$, where h is the number of simple components of $\mathfrak{g} \otimes \bar{k}$. Show that K is a product of m commutative fields, where m is the number of simple components of $\mathfrak{g}$.

   c) One says that $\mathfrak{g}$ is *absolutely simple* if $\mathfrak{g} \otimes \bar{k}$ is simple. Show that this is equivalent to $K = k$. Show that, if $\mathfrak{g}$ is simple, K is a commutative field and $\mathfrak{g}$ is absolutely simple for its natural structure of Lie algebra over K.

   d) Conversely, let K be a finite extension of k, and let $\mathfrak{g}$ be an absolutely simple Lie algebra over K. Show that $\mathfrak{g}$ is simple as a Lie algebra over k.

   e) Example: $\mathfrak{g} =$ Lie algebra of the orthogonal group of a quadratic form in 4 variables, with discriminant d not a square. Show that K is the quadratic extension $k(\sqrt{d})$.

6. Let G be a complex connected Lie group, let K be a real group-submanifold of G, and let $\mathfrak{g}$ and $\mathfrak{k}$ be the corresponding Lie algebras ($\mathfrak{g}$ is over $\mathbf{C}$, and $\mathfrak{k}$ over $\mathbf{R}$).

   a) Assume $\mathfrak{k} + i\mathfrak{k} = \mathfrak{g}$. Show that any complex group submanifold of G containing K is equal to G itself.

   b) Show that (a) is satisfied in the following cases:

   (i) $G = \mathrm{SL}(n, \mathbf{C})$, $K = \mathrm{SU}(n) =$ special unitary group

   (ii) $G = \mathrm{SO}(n, \mathbf{C})$, $K = \mathrm{SO}(n) =$ special real orthogonal group

   (iii) $G = \mathrm{Sp}(2n, \mathbf{C})$, $K = \mathrm{SU}(2n) \cap G =$ quatern. unitary group.

7. Assume k is algebraically closed; let $\mathfrak{g}_i$ ($i = 1, 2$) be Lie algebras over k and let $\mathfrak{g} = \mathfrak{g}_1 \times \mathfrak{g}_2$.

   a) Let V_i be an irreducible $\mathfrak{g}_i$-module. Show that $V_1 \otimes V_2$ is an irreducible $\mathfrak{g}$-module.

   b) Show that any irreducible $\mathfrak{g}$-module is isomorphic to some $V_1 \otimes V_2$ as above.

   c) What happens when k is not algebraically closed?

8. Let $\mathfrak{g}$ be a real Lie algebra whose Killing form is positive definite. Show that $\mathfrak{g} = 0$.

Chapter VII. Representations of $\mathfrak{sl}_n$

In this chapter, k denotes an algebraically closed field of characteristic zero. All Lie algebras and all modules are supposed to be finite-dimensional over k.

1. Notations

Let n be an integer ≥ 2, and let $\mathfrak{g} = \mathfrak{sl}_n$: Lie algebra of $n\times n$-matrices $x = (x_{ij})$ with $\mathrm{Tr}(x) = 0$. Since the center of $\mathfrak{g}$ is zero, and k^n is an irreducible $\mathfrak{g}$-module, $\mathfrak{g}$ is *semisimple* (cf. Chapter VI – one could also check this by computing the Killing form of $\mathfrak{g}$). In fact, $\mathfrak{g}$ is even simple (cf. Exer. 1), but we will not have to use this.

Define now

$\mathfrak{h}$ = Lie algebra of diagonal matrices $H = (\lambda_1, \ldots, \lambda_n)$ with $\sum \lambda_i = 0$.
$\mathfrak{n}_+$ = Lie algebra of strictly superdiagonal matrices (i.e., matrices (a_{ij}) with $x_{ij} = 0$ for $i \geq j$).
$\mathfrak{n}_-$ = Lie algebra of strictly infradiagonal matrices.

This gives a direct sum decomposition of $\mathfrak{g}$ (as a vector space):

$$\mathfrak{g} = \mathfrak{n}_- \oplus \mathfrak{h} \oplus \mathfrak{n}_+ \, .$$

Note that $\mathfrak{h}$ is abelian, $\mathfrak{n}_+$ and $\mathfrak{n}_-$ are nilpotent (cf. V.2). For $n = 2$, one has:

$$\mathfrak{h} = \begin{pmatrix} * & 0 \\ 0 & -* \end{pmatrix}, \quad \mathfrak{n}_+ = \begin{pmatrix} 0 & * \\ 0 & 0 \end{pmatrix}, \quad \mathfrak{n}_- = \begin{pmatrix} 0 & 0 \\ * & 0 \end{pmatrix}.$$

We also put $\mathfrak{b} = \mathfrak{h} \oplus \mathfrak{n}_+$; $\mathfrak{b}$ is a solvable subalgebra of $\mathfrak{g}$ (the canonical "Borel subalgebra"); its derived algebra $[\mathfrak{b}, \mathfrak{b}]$ is $\mathfrak{n}_+$.

Let $\mathfrak{h}^*$ be the dual of $\mathfrak{h}$. An element $\chi \in \mathfrak{h}^*$ can be written

$$\chi(H) = u_1\lambda_1 + \cdots + u_n\lambda_n \, , \qquad u_i \in k, \text{ if } H = (\lambda_1, \ldots, \lambda_n).$$

Since $\sum \lambda_i = 0$, the u_i's are only determined up to the addition of a constant.

Let R_+ be the subset of $\mathfrak{h}^*$ made of the linear forms $\lambda_i - \lambda_j$ $(i < j)$ and let $R = R_+ \cup (-R_+)$. An element α of R (resp. R_+) is called a *root* (resp. a *positive root*). The positive roots:

$$\alpha_1 = \lambda_1 - \lambda_2,\ \alpha_2 = \lambda_2 - \lambda_3,\ \ldots,\ \alpha_{n-1} = \lambda_{n-1} - \lambda_n$$

are called the *fundamental roots.* Any positive root $\alpha = \lambda_i - \lambda_j$ $(i < j)$ can be written as a sum of fundamental roots:

$$\alpha = \alpha_i + \alpha_{i+1} + \cdots + \alpha_{j-1} \, .$$

Let $\alpha = \lambda_i - \lambda_j$ $(i \neq j)$ be a root. Define elements H_α and X_α of $\mathfrak{g}$ in the following way:

X_α = matrix with (i,j)-entry equal 1, and all other entries zero.
H_α = element of $\mathfrak{h}$ whose i^{th} coordinate is 1, j^{th} coordinate -1, and others zero.

Note that $\alpha(H_\alpha) = 2$.

Proposition 1.1. (a) *The X_α's ($\alpha \in R_+$) make a basis of $\mathfrak{n}_+$; the $X_{-\alpha}$'s ($\alpha \in R_+$) make a basis of $\mathfrak{n}_-$.*
(b) *If $H \in \mathfrak{h}$, $\alpha \in R$, $[H, X_\alpha] = \alpha(H)X_\alpha$.*
(c) $[X_\alpha, X_{-\alpha}] = H_\alpha$.

Assertion (a) is clear. To prove (b), let $(\lambda_1, \ldots, \lambda_n)$ be the diagonal terms of H; if α is the linear form $\lambda_i - \lambda_j$, one finds that $H \cdot X_\alpha = \lambda_i \cdot X_\alpha$ and $X_\alpha \cdot H = \lambda_j \cdot X_\alpha$. Hence $[H, X_\alpha] = (\lambda_i - \lambda_j)X_\alpha = \alpha(H)X_\alpha$. A similar computation proves (c).

Example 1.2. For $n = 2$, there is just one positive root $\alpha = \lambda_1 - \lambda_2$. The elements

$$H_\alpha = \begin{pmatrix} 1 & 0 \\ 0 & -1 \end{pmatrix}, \quad X_\alpha = \begin{pmatrix} 0 & 1 \\ 0 & 0 \end{pmatrix}, \quad X_{-\alpha} = \begin{pmatrix} 0 & 0 \\ 1 & 0 \end{pmatrix}$$

make a basis of $\mathfrak{sl}_2$.

2. Weights and primitive elements

Let V be a $\mathfrak{g}$-module. If $\chi \in \mathfrak{h}^*$, we denote by V_χ the vector space of elements $v \in V$ such that $H \cdot v = \chi(H) \cdot v$ for all $H \in \mathfrak{h}$; such a v is called an *eigenvector* of $\mathfrak{h}$ of *weight* χ.

Proposition 2.1. *If $\alpha \in R$ and $v \in V_\chi$, then $X_\alpha \cdot v \in V_{\chi+\alpha}$.*

Indeed

$$\begin{aligned} HX_\alpha v &= [H, X_\alpha]v + X_\alpha Hv \\ &= \alpha(H)X_\alpha v + X(H)X_\alpha v \\ &= (\chi + \alpha)(H)X_\alpha v \ , \end{aligned}$$

hence $X_\alpha v$ is of weight $\chi + \alpha$.

Proposition 2.2. *V is the direct sum of the V_χ's (for $\chi \in \mathfrak{h}^*$).*

It is well known that non-zero eigenvectors of distinct eigenvalues are linearly independent. Hence, the sum $W = \sum_{\chi \in \mathfrak{h}^*} V_\chi$ is a direct sum. Proposition 2.1 shows that W is stable by the X_α's; since it is also stable by $\mathfrak{h}$, it is stable by $\mathfrak{g}$. Hence (complete reducibility!) V is the direct sum of W and another $\mathfrak{g}$-submodule V'. Suppose $V' \neq 0$. Since $\mathfrak{h}$ is abelian and k algebraically closed, there exists in V' at least one non-zero eigenvector v of $\mathfrak{h}$. Such a v is contained in some V_χ, and this contradicts the fact that $V' \cap W = 0$. Hence $V' = 0$, and $V = W$. q.e.d.

Definition 2.3. The χ's such that $V_\chi \neq 0$ are called the *weights* of V. The dimension of V_χ is called the *multiplicity* of χ.

Example 2.4. The weights of $\mathfrak{g}$ (for the adjoint representation) are the roots $\alpha \in R$, which have multiplicity one, and 0, which has multiplicity $n-1$.

Proposition 2.5. *Let $v \in V$. The following conditions are equivalent:*
(1) *v is an eigenvector for the Borel algebra $\mathfrak{b} = \mathfrak{h} \oplus \mathfrak{n}_+$.*
(2) *v is an eigenvector for $\mathfrak{h}$, and $X_\alpha v = 0$ for all $\alpha \in R_+$.*

This follows from the fact that $\mathfrak{n}_+ = [\mathfrak{b}, \mathfrak{b}]$ and that the X_α's ($\alpha \in R_+$) make a basis of $\mathfrak{n}_+$.

Definition 2.6. A non-zero element $v \in V$ which satisfies the equivalent conditions of Prop. 2.5 is called *primitive.*

Note that a primitive element has a well-defined weight $\chi \in \mathfrak{h}^*$.

Proposition 2.7. *Any non-zero $\mathfrak{g}$-module V contains a primitive element.*

This follows from Lie's theorem (cf. Chap. V) applied to the $\mathfrak{b}$-module V.

[*Alternate proof*: Let S be the set of weights of V. Using the fact that S is finite, and non-empty (cf. Prop. 2.2), one sees that S contains an element χ such that $\chi + \alpha_i \notin S$ for any i. The non-zero elements of the corresponding V_χ are primitive.]

3. Irreducible $\mathfrak{g}$-modules

Theorem 3.1. *Let V be a $\mathfrak{g}$-module, and let $v \in V$ be a primitive element of weight χ. Let $V_1 = (U\mathfrak{g}) \cdot v$ be the $\mathfrak{g}$-submodule of V generated by v. Then:*
(a) *V_1 is irreducible.*
(b) *The weights of V_1 are of the form $\chi - \sum_{i=1}^{n-1} m_i\alpha_i$, with $m_i \geq 0$.*
(c) *Any element of V_1 which is of weight χ is a multiple of v.*

The universal algebra $U\mathfrak{g}$ can be written

$$U\mathfrak{g} = U\mathfrak{n}_- \otimes U\mathfrak{b}\,, \qquad \text{cf. Chap. III.}$$

Since v is an eigenvector of $\mathfrak{b}$, $(U\mathfrak{b}) \cdot v = k \cdot v$, hence $V_1 = (U\mathfrak{g}) \cdot v$ is equal to $(U\mathfrak{n}_-) \cdot v$. Applying Birkhoff-Witt to $U\mathfrak{n}_-$, we then see that V_1 is generated by elements of the form $M \cdot v$, where M is a monomial in the $X_{-\alpha}$'s ($\alpha \in R_+$). Prop .2.1 then shows that the $M \cdot v$ are eigenvectors of $\mathfrak{h}$ of weight $\chi - \sum_{\alpha>0} q_\alpha \alpha$, with $q_\alpha \geq 0$; this implies (b). Assertion (c) follows from the fact that the q_α's can be all zero only when M is of degree zero (i.e., $M = 1$), and in that case $M \cdot v = v$. To prove (a), suppose V_1 is decomposed into the direct sum of two $\mathfrak{g}$-modules V' and V''; let $v = v' + v''$ be the corresponding decomposition of v. Since $(V_1)_\chi = V'_\chi \oplus V''_\chi$, v' and v'' are both of weight χ; (c) then shows that they are multiples of v, and one of them must be zero, say v''; hence $v' = v$, and since v generates V_1, one has $V' = V_1$, $V'' = 0$. q.e.d.

Theorem 3.2. (1) *Let V be an irreducible $\mathfrak{g}$-module. Then V contains only one primitive element (up to multiplication by a non-zero element of k); the weight of such an element is called the highest weight of V.*

(2) *Two irreducible $\mathfrak{g}$-modules with the same highest weight are isomorphic.*

If V is irreducible, it contains at least one primitive element v (cf. Prop. 2.7); let χ be the weight of v. Let now v' be another primitive element of V, and let χ' be its weight. Since V is irreducible, v generates V, hence Theorem 3.1 shows that

$$\chi - \chi' = \sum_{i=1}^{n-1} m_i \alpha_i \,, \qquad \text{with } m_i \geq 0 \text{ for all } i.$$

The same argument, applied to v', shows that:

$$\chi' - \chi = \sum_{i=1}^{n-1} m'_i \alpha_i \,, \qquad \text{with } m'_i \geq 0 \text{ for all } i.$$

These two relations imply $m_i = m'_i = 0$, i.e., $\chi = \chi'$, and part (c) of Theorem 3.1 then shows that v' is a scalar multiple of v. This proves (1).

Let now V_1, V_2 be two irreducible $\mathfrak{g}$ modules having primitive elements v_1, v_2 of the same weight χ. The element $v = (v_1, v_2)$ of $V_1 \oplus V_2$ is again primitive of weight χ. By Theorem 3.1, the $\mathfrak{g}$-submodule W of $V_1 \times V_2$ generated by v is irreducible. The projection map $\pi_i : W \to V_i$ is non-zero (since $\pi_i(v) = v_i$), hence is an isomorphism, W and V_i being irreducible. This shows that V_1 and V_2 are both isomorphic to W.

Remark. Theorem 3.2 reduces the classification of irreducible $\mathfrak{g}$-modules to the determination of the elements $\chi \in \mathfrak{h}^*$ which are "highest weights", i.e., weights of primitive elements in some $\mathfrak{g}$-module. This determination will be made in VII.4.

4. Determination of the highest weights

Theorem 4.1. *Let χ be an element of $\mathfrak{h}^*$, and write χ in the form:*

$$\chi(\lambda_1, \ldots, \lambda_n) = u_1\lambda_1 + \cdots + u_n\lambda_n \,.$$

There exists an irreducible $\mathfrak{g}$*-module with highest weight* χ *if and only if* $u_i - u_j$ *is a positive integer for all* $i < j$.

(Of course, "positive" means ≥ 0.)

Proof of necessity. Note first that $u_i - u_j = \chi(H_\alpha)$ if α is the positive root $\lambda_i - \lambda_j$. Hence, we have to prove that $\chi(H_\alpha)$ is a positive integer (for $\alpha \in R_+$) if χ is the weight of a primitive element v.

Proposition 4.2. *Let* v *be a primitive element of weight* χ*, and define* $v_m^\alpha = (X_{-\alpha})^m v/m!$ *for* $m \geq 0$*, where* $(X_{-\alpha})^m$ *is the* m^{th} *iterate of* $X_{-\alpha}$*. Then, the following formulas hold:*

(i) $X_{-\alpha} v_m^\alpha = (m+1) v_{m+1}^\alpha$

(ii) $H v_m^\alpha = (\chi - m\alpha)(H) v_m^\alpha$ *if* $H \in \mathfrak{h}$

(iii) $X_\alpha v_m^\alpha = (\chi(H_\alpha) - m + 1) v_{m-1}^\alpha$.

Formula (i) is obvious; formula (ii) means that v_m^α is of weight $\chi - m\alpha$, which follows from Prop. 2.1. One proves (iii) by induction on m, the case $m = 0$ being trivial (it is understood that $v_{-1}^\alpha = 0$ – note that this convention agrees with (i) for $m = -1$). If $m \geq 1$, one writes:

$$\begin{aligned} m \cdot X_\alpha v_m^\alpha &= X_\alpha X_{-\alpha} v_{m-1}^\alpha = H_\alpha v_{m-1}^\alpha + X_{-\alpha} X_\alpha v_{m-1}^\alpha \\ &= \lambda \cdot v_{m-1}^\alpha , \end{aligned}$$

with $\lambda = \chi(H_\alpha) - (m-1)\alpha(H_\alpha) + (m-1)(\chi(H_\alpha) - m + 2)$.

Using the fact that $\alpha(H_\alpha) = 2$, one sees that $\lambda = m(\chi(H_\alpha) - m + 1)$, and this proves (iii).

Corollary 4.3. *There exists* $m \geq 0$ *such that* $v_m^\alpha \neq 0$ *and* $v_{m+1}^\alpha = 0$*. One has* $\chi(H_\alpha) = m$.

Since the v_m^α's have weights $\chi - m\alpha$, and the number of possible weights of a given module is finite, one must have $v_m^\alpha = 0$ for large m, hence the existence of m with $v_m^\alpha \neq 0$, $v_{m+1}^\alpha = 0$. Applying formula (iii) for $m+1$, one finds:

$$0 = X_\alpha v_{m+1}^\alpha = (\chi(H_\alpha) - m) v_m^\alpha .$$

Since $v_m^\alpha \neq 0$, this implies $\chi(H_\alpha) = m$.

This finishes the proof of the "necessity" part.

Proof of sufficiency. Let $\pi_1, \ldots, \pi_{n-1}$ be the linear forms $\lambda_1, \lambda_1 + \lambda_2, \ldots, \lambda_1 + \cdots + \lambda_{n-1}$. The condition of Theorem 4.1 is equivalent to the following:

$$\chi \text{ can be written } \chi = \sum_{i=1}^{n-1} m_i \pi_i ,$$

where the m_i's are positive integers.

Proposition 4.4. *If χ and χ' are the highest weights of the irreducible modules V and V', $\chi + \chi'$ is the highest weight of an irreducible submodule of $V \otimes V'$.*

Let v and v' be primitive elements of V and V'. Then $v \otimes v'$ is a primitive element of $V \otimes V'$ and its weight is $\chi + \chi'$. The $\mathfrak{g}$-submodule W generated by $v \otimes v'$ is irreducible (Theorem 3.1), and its highest weight is $\chi + \chi'$.

Corollary 4.5. *The set of highest weights is closed under addition.*

Hence, to prove that χ is a highest weight, it is enough to prove that the π_i's are highest weights. We do this by giving explicitly the corresponding irreducible $\mathfrak{g}$-modules:

Proposition 4.6. *Let V be k^n, viewed in a natural way as a $\mathfrak{g}$-module. For $1 \leq i \leq n-1$, let V_i be the exterior i^{th}-power of V. Then V_i is an irreducible $\mathfrak{g}$-module of highest weight π_i.*

Let $e_1, \ldots, e_n$ be the canonical basis of V, and let $v_i = e_1 \wedge \cdots \wedge e_i$. A simple computation shows that v_i is a primitive element of V_i, of weight π_i; moreover, by applying to v_i a monomial in the $X_{-\alpha}$'s, one can obtain any term of the form $e_{m_1} \wedge \cdots \wedge e_{m_i}$, $m_1 < \cdots < m_i$; hence V_i is irreducible (Theorem 3.1). This concludes the proof of Theorem 4.1.

Remarks. 1. Analogous results are true for any semisimple Lie algebra. In fact, all the proofs we have given (except the last one – based on an explicit construction of irreducible modules) apply to the general case, once the fundamental properties of "Cartan subalgebras" and "roots" have been proved.

2. Theorem 4.1 shows that the classes of irreducible $\mathfrak{g}$-modules are in one-to-one correspondence with systems $(m_1, \ldots, m_{n-1})$ of $n-1$ positive integers. For an explicit description of the module which corresponds to $(m_1, \ldots, m_{n-1})$, see for instance H. Weyl, *The Classical Groups*, Chapter IV.

3. When $n = 2$, there is just one integer m, and the corresponding irreducible module is the mth-symmetric power of $V = k^2$.

Exercises

1. Suppose $\mathfrak{g} = \mathfrak{sl}_n$ is the product of two semisimple Lie algebras $\mathfrak{g}_1$ and $\mathfrak{g}_2$. Prove that the $\mathfrak{g}$-module $k^n = V$ is a tensor product $V_1 \otimes V_2$, where V_i is an irreducible $\mathfrak{g}_i$-module (cf. Chap. VI). If $n_i = \dim V_i$, one has $n = n_1 n_2$, $\dim \mathfrak{g}_i \leq n_i^2 - 1$. Show that this implies that one of the n_i's is equal to 1, hence $\mathfrak{g}_i = 0$, and $\mathfrak{g}$ is simple.

2. Show that all the results of this Chapter hold when k is an arbitrary field of characteristic zero (Hint: use the fact that, over an algebraic closure $\overline{k}$ of k, all weights take rational values on the H_α's; this is enough to imply that Prop. 2.2 and 2.7 hold over k; the rest offers no difficulty.)
3. Let $k = \mathbf{C}$, field of complex numbers. The group $G = \mathrm{SL}(n, \mathbf{C})$ contains the subgroup $\mathrm{SU}(n)$ of unitary matrices with det = 1. Show that $G/\mathrm{SU}(n)$ is homeomorphic to an Euclidean space $\mathbf{R}^N$ (Hint: identity this homogeneous space with the space of all positive definite hermitian forms on $\mathbf{C}^n$). Show that $\mathrm{SU}(n)/\mathrm{SU}(n-1)$ is homeomorphic to the sphere $\mathbf{S}_{2n-1}$. Use this fact to prove (by induction on $n \geq 2$) that $\mathrm{SU}(n)$ and G are connected and simply connected. Hence, any linear representation of $\mathfrak{g} = L(G)$ corresponds to an analytic representation of G, and conversely.
4. Same notations and assumptions as in Exer. 3. Show that the subalgebra $\mathfrak{h}$ of $\mathfrak{g}$ corresponds to a group submanifold of G which is isomorphic to a product of $n-1$ copies of $\mathbf{C}^* = \mathbf{C}/\mathbf{Z}$. Use this to give a direct proof of the fact that any weight of $\mathfrak{g}$ is a linear combination (with integral coefficients) of the π_i's.
5. a) Let P (resp. Q) be the subgroup of $\mathfrak{h}^*$ generated by the π_i's (resp. by the roots). Define an exact sequence:

$$0 \longrightarrow Q \xrightarrow{i} P \xrightarrow{e} \mathbf{Z}/n\mathbf{Z} \longrightarrow 0 ,$$

where i is the inclusion map, and $e(\pi_i) = i$ for $1 \leq i \leq n-1$.

b) Let V be an irreducible $\mathfrak{g}$-module. Show that all weights of V are elements of P, and have the same image by e; let $e(V) \in \mathbf{Z}/n\mathbf{Z}$ be this image.

c) Assume $k = \mathbf{C}$ (cf. Exer. 3). Prove that the center C of $G = \mathrm{SL}(n, \mathbf{C})$ is a cyclic group of order n, made of the scalar matrices w with $w^n = 1$. Let V be an irreducible $\mathfrak{g}$-module; show that the image of $w \in C$ by the corresponding representation of G is a scalar which is equal to $w^{e(V)}$.

d) Using (c), prove that the irreducible representations of the projective group $\mathrm{PGL}(n, \mathbf{C}) = G/C$ correspond to the irreducible $\mathfrak{g}$-modules V with $e(V) = 0$.
6. Let χ be any element of $\mathfrak{h}^*$. Let L_χ be a one-dimensional $\mathfrak{b}$-module of weight χ. Let $E_\chi = L_\chi \otimes_{U\mathfrak{b}} U\mathfrak{g}$ be the corresponding "induced $\mathfrak{g}$-module" – which is infinite-dimensional. Show that E_χ contains a primitive element v of weight χ. What are the other weights of E_χ? Show that there is a largest submodule H of E_χ which does not contain v. The quotient $V_\chi = E_\chi/H$ is irreducible; show that it is finite dimensional if and only if χ satisfies the conditions of Theorem 4.1. Give an explicit description of V_χ when $n = 2$.
7. Let $n = 4$, and let V be the irreducible $\mathfrak{g}$-module with highest weight π_2 (cf. Prop. 4.6). Show that $\dim V = 6$, and that there is a non-degenerate invariant quadratic form on V. Use this to construct an isomorphism of $\mathfrak{sl}_4$ onto the Lie algebra of the orthogonal group in 6 variables.

Part II – Lie Groups

Introduction

This part is meant as an introduction to formal groups, analytic groups, and the correspondence between them and Lie algebras (Lie theory). Analytic groups are defined over any complete field (real, complex or ultrametric); Lie theory applies equally well to these cases, provided the characteristic is zero.

I have made an essential use of unpublished manuscripts of N. Bourbaki, both on analytic manifolds, and on Lie groups.

Part II has been written by R. Rasala. I want to thank him for the good job he has done; many improvements on the oral exposition are due to him.

Jean-Pierre Serre

Harvard, Fall 1964

Chapter I. Complete Fields

Definition. Let k be a commutative field. An *absolute value* on k is a function $k \to \mathbf{R}_+$ denoted by $x \mapsto |x|$, $x \in k$ satisfying the conditions:

1. $|x| = 0 \iff x = 0$.
2. $|xy| = |x|\,|y|$.
3. $|1| = 1$.
4. $|x + y| \leq |x| + |y|$.

Examples.
(i) Define

$$\begin{cases} |x| = 0 & \text{if } x = 0 \\ |x| = 1 & \text{if } x \neq 0 \end{cases}.$$

The topology on k induced by this absolute value is discrete.

From now on we will assume that the absolute value is non-trivial, i.e., there exists $x \in k$ with $0 < |x| < 1$.

(ii) $\mathbf{R}$, $\mathbf{C}$ with the usual absolute values.

(iii) If we replace condition 4) by 4′) $|x-y| \leq \sup\{|x|, |y|\}$, such an absolute value is called ultrametric or non-archimedian.

Note. The condition 4′) is equivalent to the following:
For any $\varepsilon \geq 0$, the relation $|x - y| \leq \varepsilon$ is an equivalence relation.

Now suppose k is complete for an ultrametric absolute value.

Theorem. *Let $\{x_n\}$ be a sequence with $x_n \in k$. Then $\sum x_n$ converges if and only if $x_n \to 0$.*

The proof is immediate.

Theorem (Ostrowski). *Let k be a complete field for an absolute value. Then either $k = \mathbf{R}$ or $\mathbf{C}$ with the usual absolute value $|x|^\alpha$, $0 < \alpha \leq 1$ or the absolute value of k is ultrametric.*

For the proof, see for instance, Bourbaki, *Alg. Comm.*, Chap. VI, §6.

Let k again be a field with an ultrametric absolute value $|x|$ and let a be a real number with $0 < a < 1$. Define a real number $v(x)$ by the formula $|x| = a^{v(x)}$. Then $v(x)$ satisfies the conditions:

1. $v(x) = +\infty \iff x = 0$.
2. $v(xy) = v(x) + v(y)$.

3. $v(1) = 0$.

4. $v(x + y) \geq \inf(v(x), v(y))$.

The number $v(x)$ is called the *valuation* of x.

Examples.

a. Let $k = \mathbf{C}((T))$ be the field of the power series in one variable T.

Let $a = \sum_{n \gg -\infty} a_n T^n$, $a_n \in \mathbf{C}$, $a_n = 0$ for $-n$ large enough.

Define $v(a) =$ smallest n such that $a_n \neq 0$. Then $a = T^n(\alpha_0 + \alpha_1 T + \cdots$ with $\alpha_0 \neq 0$, $n \in \mathbf{Z}$, i.e., $v(a) \geq n \iff a = T^n \varphi$ with $\varphi \in \mathbf{C}[[T]]$.

Note that the field $\mathbf{C}((T))$ is complete.

b. Let $\mathbf{Q}$ be the field of rational numbers and choose a prime number p. For any $a \in \mathbf{Q}$, $a \neq 0$, we write $a = \frac{r}{s} = p^n \frac{r'}{s'}$ where r', s' are integers prime to p.

The valuation defined by $v(a) = n$ is called the *p-adic valuation* of a.

The p-adic completion of $\mathbf{Q}$ is denoted by $\mathbf{Q}_p$ and called the field of p-adic numbers.

Note that $a_n \to 0$ in the p-adic sense if and only if a_n is divisible by a power of p, say p^{h_n}, where $h_n \to \infty$.

Definition. Let k be a field and let v be a valuation of k. Then the set

$$A_v = \{\, x \mid x \in k,\ v(x) \geq 0 \,\}$$

is a ring and it is called the *ring of the valuation* v.

Example. Let $k = \mathbf{Q}_p$ and let v be the p-adic valuation; then $A_v = \mathbf{Z}_p$ is the ring of p-adic integers.

If $\alpha \geq 0$ is a real number, then the sets

$$I_\alpha = \{\, x \mid x \in A_v,\ v(x) \geq \alpha \,\}$$
$$I'_\alpha = \{\, x \mid x \in A_v,\ v(x) > \alpha \,\}$$

are ideals of A_v.

In particular, if $\alpha = 0$ we have

$$I'_0 = \mathfrak{m}_v = \{\, x \mid v(x) > 0 \,\}$$

which is the maximal ideal of A_v. The field $k(v) = A_v/\mathfrak{m}_v$ is called the residue field of v.

Examples.

(i) Let $k = \mathbf{C}((T))$, then

$$A_v = \mathbf{C}[[T]]$$
$$\mathfrak{m}_v = (T)$$

and $k_v = A_v/\mathfrak{m}_v = \mathbf{C}$.

(ii) Let $k = \mathbf{Q}_p$, then

$$A_v = \mathbf{Z}_p$$
$$\mathfrak{m}_v = (p) = p\mathbf{Z}_p$$

and $k_v = \mathbf{F}_p = \mathbf{Z}/p\mathbf{Z}$.

Theorem. *Let $x \mapsto |x|$ be an ultrametric absolute value on $\mathbf{Q}$. Then either*

$$|x| = 1 \qquad \textit{for all } x \neq 0$$

or the absolute value $|\ |$ is the p-adic value for some p.

Proof. Suppose that there exists a rational number $r \in \mathbf{Q}$ such that

$$0 < |r| < 1\ ;$$

this implies that there exists a prime number p such that $|p| \neq 1$, i.e., $0 < |p| < 1$ (notice that if $n \in \mathbf{Z}$ then $|n| \leq 1$).

Let $n \in \mathbf{Z}$ and assume $(n,p) = 1$, then there exist $A, B \in \mathbf{Z}$ with $An + Bp = 1$; in the case $|n| < 1$ we get $|An| < 1$, but we know that $|p| < 1$, i.e., $|Bp| < 1$ which implies $|1| < 1$, so we must have $|n| = 1$.

Now take $a = |p|$ and $r = p^{v_p(r)} \frac{n}{n'}$, where $n, n' \in \mathbf{Z}$ are prime to p. We have $|r| = a^{v_p(r)}$, q.e.d.

Corollary. *If k is a complete field with respect to an ultrametric absolute value $|\ |$ and the characteristic of k is zero then*

$$k \supset \mathbf{Q} \textit{ with the discrete topology}$$

or

$$k \supset \mathbf{Q}_p \textit{ for } |\ |.$$

Chapter II. Analytic Functions

We first fix some notation:

1. k: field, complete with respect to a non-trivial absolute value.
   $k[[X_1,\dots,X_n]]$: formal power series in n variables $X_1,\dots,X_n$.

2. We use:
   a. Greek letters α, β for n-tuples as $\alpha=(\alpha_1,\dots,\alpha_n)$, $\alpha_i\ge 0$, $\in \mathbf{Z}$.
   b. Latin letters r, s for n-tuples as $r=(r_1,\dots,r_n)$, $r_i>0$, $\in \mathbf{R}$.
   c. Latin letters x, y for n-tuples as $x=(x_1,\dots,x_n)$, $x_i\in k$.

3. We set:
   $r^\alpha = r_1^{\alpha_1}\cdots r_n^{\alpha_n}$
   $x^\alpha = x_1^{\alpha_1}\cdots x_n^{\alpha_n}$
   $X^\alpha = X_1^{\alpha_1}\cdots X_n^{\alpha_n}$
   $|\alpha| = \sum \alpha_i$
   $\alpha! = \prod \alpha_i!$
   $\binom{\alpha}{\beta} = \frac{\alpha!}{\beta!(\alpha-\beta)!}$

4. We define:
   $|x|\le r$ (resp. $|x|<r$) $\iff$ $|x_i|\le r_i$ (resp. $|x_i|<r_i$), $1\le i\le n$.
   We define similarly $r'\le r$, $r'<r$, $\alpha'\le\alpha$, and $\alpha'<\alpha$.

5. We set:
   $P(r)(x)=\{\,y:|y-x|\le r\,\}$ = Polydisk of radius r about x.
   $P_0(r)(x)=\{\,y:|y-x|< r\,\}$ = Strict polydisk of radius r about x.
   $P(r)=P(r)(0)$
   $P_0(r)=P_0(r)(0)$.

Definition. Let $f=\sum a_\alpha X^\alpha$ and let r be as above.

1. The series f is said to be *convergent in* $P(r)$ if

$$\sum |a_\alpha| r^\alpha < \infty\,. \tag{1}$$

2. The series f is said to be *convergent in* $P_0(r)$ if it is convergent in $P(r')$ for all $r'<r$.

Lemma. *Let $f=\sum a_\alpha X^\alpha$ and let r be as above. Then:*

1. *If f converges in $P(r)$, there is a constant M such that for all α*

$$|a_\alpha| r^\alpha \le M\,. \tag{2}$$

2. *If there is a constant M such that (2) holds for all α, f converges in $P_0(r)$ and uniformly in $P(r')$ for $r'<r$.*

Proof.
1. Take $M=\sum |a_\alpha| r^\alpha$ which is finite by (1).

2. Suppose $r' < r$. Then:

$$\begin{aligned}\sum |a_\alpha| r'^\alpha &= \sum (|a_\alpha| r^\alpha) \frac{r'^\alpha}{r^\alpha} \\ &\le M \sum \frac{r'^\alpha}{r^\alpha} \\ &= M \prod \left(1 - \frac{r'_i}{r_i}\right)^{-1}.\end{aligned}$$

This shows that f converges uniformly in $P(r')$. In particular, f converges in $P_0(r)$.

Definition. Let $f = \sum a_\alpha X^\alpha$. The series f is said to be *convergent* if it is convergent in $P_0(r)$ for some $r > 0$.

Let $f = \sum a_\alpha X^\alpha$ be convergent in $P_0(r)$. For any $x \in P_0(r)$, the series $\sum a_\alpha x^\alpha$ converges absolutely (and uniformly in any $P(r')$ for $r' < r$); its sum $\tilde{f}(x)$ is a continuous function of x. We have the following lemma:

Lemma. $\tilde{f} = 0 \Rightarrow f = 0$.

Proof.
$n = 1$:

Suppose $f \neq 0$. Then

$$f(X) = X^n(c_0 + c_1 X + \cdots), \qquad c_0 \neq 0,\ m \ge 0.$$

The series $\sum c_i X^i$ is convergent. The function this series defines is non-zero at 0 and hence non-zero in a neighborhood U of 0 by continuity. Now X^m is non-zero in $U - \{0\}$ so that $\tilde{f}$ does not vanish identically on U. In fact if $m > 0$, the zero of $\tilde{f}$ at 0 is isolated.

$n > 1$:

We assume the lemma for $n - 1$ and suppose $\tilde{f} = 0$. Write

$$f = \sum_i c_i(X_1, \ldots, X_{n-1}) X_n^i, \qquad c_i \in k[[X_1, \ldots, X_{n-1}]].$$

Since f is convergent in $P_0(r)$, the c_i are convergent in the $(n-1)$-dimensional polydisk $P_0(s)$ where $s = (r_1, \ldots, r_{n-1})$.

By hypothesis, for $y = (y_1, \ldots, y_{n-1}) \in P_0(s)$, the function g defined by

$$g(x_n) = \sum_i \tilde{c}_i(y_1, \ldots, y_{n-1}) x_n^i$$

is identically zero. Hence, by the case $n = 1$, all $\tilde{c}_i(y_1, \ldots, y_{n-1})$ are 0. Since this is true for all $y \in P_0(s)$, all c_i are 0 by induction. Hence, $f = 0$.

By the above lemma, we may identify a convergent power series f with its associated function $\tilde{f}$.

We shall now study analytic functions.

Definition. Let $U \subset k^n$ be open and let $\phi : U \to k$ be a function. Then ϕ is said to be *analytic* in U if for each $x \in U$ there is a formal power series f and a radius $r > 0$ such that:

1. $P_0(r)(x) \subset U$.
2. f converges in $P_0(r)$ and, for $h \in P_0(r)$, $\phi(x+h) = f(h)$.

Remark. If ϕ is analytic in U and $x \in U$, then the power series f of the above definition is unique and is called the *local expansion* of ϕ at x.

Theorem. *Let $f = \sum a_\alpha X^\alpha$ be convergent in $P_0(r)$ for $r > 0$. Then f is an analytic function in $P_0(r)$.*

Proof. Let $x \in P_0(r)$. Then we may choose r' such that $|x| \le r' < r$. Set $s = r - r'$. Next note that:

$$(x+h)^\alpha = \sum_{\beta \le \alpha} \binom{\alpha}{\beta} x^{\alpha-\beta} h^\beta \,.$$

Hence:

$$f(x+h) = \sum_\alpha a_\alpha \left(\sum_{\beta \le \alpha} \binom{\alpha}{\beta} x^{\alpha-\beta} h^\beta \right) , \qquad h \in P_0(s).$$

To show that rearrangement of the above sum is permissible, we shall show that:

$$(*) \qquad \sum_\alpha \sum_{\beta \le \alpha} \left| a_\alpha \binom{\alpha}{\beta} x^{\alpha-\beta} h^\beta \right| < \infty \,, \qquad h \in P_0(s).$$

Indeed, let $|h| \le s' < s$. Then:

$$\left| a_\alpha \binom{\alpha}{\beta} x^{\alpha-\beta} h^\beta \right| \le |a_\alpha| \binom{\alpha}{\beta} |x|^{\alpha-\beta} |h|^\beta \,.$$

We have used the fact that $|p| \le p$ when p is a positive integer to estimate $|\binom{\alpha}{\beta}|$. Thus:

$$\sum_{\beta \le \alpha} \left| a_\alpha \binom{\alpha}{\beta} x^{\alpha-\beta} h^\beta \right| \le \sum_{\beta \le \alpha} |a_\alpha| \binom{\alpha}{\beta} r'^{\alpha-\beta} s'^\beta = |a_\alpha| (r'+s')^\alpha \,.$$

Hence an upper estimate for the sum $(*)$ is:

$$\sum_\alpha |a_\alpha| (r'+s')^\alpha < \infty \,.$$

This sum is finite since f converges in $P_0(r)$ and $r' + s' < r$.

What we have shown is the following lemma:

Lemma. *Let $f = \sum a_\alpha X^\alpha$ be convergent in $P_0(r)$ and define*

$$\Delta^\beta f = \sum_{\alpha \geq \beta} a_\alpha \binom{\alpha}{\beta} X^{\alpha-\beta} .$$

Then:

1. *$\Delta^\beta f$ is convergent in $P_0(r)$.*
2. *For $x \in P_0(r)$, the series $\sum_\beta \Delta^\beta f(x) h^\beta$ converges in $P_0(r - |x|)$.*
3. *For $x \in P_0(r)$ and $h \in P_0(r - |x|)$:*

$$f(x+h) = \sum_\beta \Delta^\beta f(x) h^\beta .$$

Subproof: Indeed, 1 and 2 follow from (∗) immediately. 3 also follows from (∗) since (∗) implies that:

$$\sum_\alpha a_\alpha \left(\sum_{\beta \leq \alpha} \binom{\alpha}{\beta} x^{\alpha-\beta} h^\beta \right) = \sum_\beta \left(\sum_{\alpha \leq \beta} a_\alpha \binom{\alpha}{\beta} x^{\alpha-\beta} \right) h^\beta = \sum_\beta \Delta^\beta f(x) h^\beta .$$

This completes the proof of the lemma and clearly also the proof of the theorem.

We now generalize the notion of analytic function to vector-valued functions.

Definition. Let $U \subset k^m$ be open and let $\phi = (\phi_1, \cdots, \phi_n) : U \to k^n$. Then ϕ is said to be *analytic* if ϕ_i is analytic for $1 \leq i \leq n$.

The lemma of the preceding theorem is a special case of the following theorem:

Theorem. *Suppose $U \xrightarrow{f} V$ and $V \xrightarrow{g} W$ are analytic where $U \subset k^m$, $V \subset k^n$, and $W \subset k^p$ are open. Then $g \circ f$ is analytic.*

Proof. We must check that $g \circ f$ has a local power series expansion about each point $x \in U$. Now, the lemma of the preceding theorem shows that the class of analytic functions defined on open subsets of some k^μ with values in some k^ν is invariant under translation of domain or range. Hence, we may assume that $x = 0$, $f(0) = 0$, $g(0) = 0$. Further, it follows from the definition of vector valued analytic functions that it suffices to consider the case $p = 1$.

Let $\sum_{\beta>0} b_\beta Y^\beta$ be a local expansion of g at 0 valid in $P_0(s)$ where $s = (s_1, \ldots, s_n)$. Let $f = (f_1, \ldots, f_n)$ and let $\sum_{\alpha>0} a_{i,\alpha} X^\alpha$ be a local expansion of f_j. We may choose $r = (r_1, \ldots, r_m)$ so that:

$$\sum_{\alpha>0} |a_{i,\alpha}| r^\alpha < \frac{s_i}{2}, \qquad 1 \le i \le n.$$

Then, for $h \in P_0(r)$:

$$g \circ f(h) = \sum_{\beta>0} b_\beta \left(\ldots, \sum_{\alpha>0} a_{i,\alpha} h^\alpha, \ldots \right)^\beta$$

To complete the proof, we must show that the right hand side defines a power series in h convergent in $P_0(r)$. Now, the right hand side does define a *formal* power series in h since formally there are only finitely many terms which contribute to coefficients of any h^α. Indeed, terms where $|\beta| > |\alpha|$ make no contribution (since all $a_{i,0} = 0$). Hence, it remains to check the convergence of the formal power series we obtain. This follows since:

$$\sum_{\beta>0} |b_\beta| \left(\ldots, \sum_{\alpha>0} |a_{i,\alpha}|\, |h|^\alpha, \ldots \right)^\beta \le \sum_{\beta>0} |b_\beta| \left(\frac{s}{2}\right)^\beta < \infty .$$

The proof of the theorem is therefore complete.

Remark. 1) The reader may consult Bourbaki, *Alg.*, Chap. IV, §4, for a more detailed discussion of the step on the existence of a formal power series.

2) There is a general method based on the theorem of Ostrowski which is useful for proving theorems such as the preceding one. One simply observes that it suffices to consider two cases:

1. $k = \mathbf{R}$ or $\mathbf{C}$.
2. k ultrametric.

Let us illustrate this method by giving an alternate proof of the above theorem:

Case 1: $k = \mathbf{R}$ or $\mathbf{C}$.

a. $k = \mathbf{C}$:

It is known that:

$$\phi \text{ is analytic} \iff \phi \text{ is } C^1 \text{ and } D\phi \text{ is a } complex \text{ linear map.}$$

Since the composite of C^1 maps is C^1, the composite of derivatives is the derivative of the composite, and the composite of complex linear maps is complex linear, the theorem follows in this case.

b. $k = \mathbf{R}$:

We may locally extend real analytic functions to complex analytic by power series. Hence the theorem in the real case follows from the complex case.

Case 2: k ultrametric.

We assume as in the original proof of the theorem that we are seeking a power series expansion of $g \circ f$ about $x = 0$ and that $f(0) = 0$ and $g(0) = 0$. We note that without loss of generality, we may replace $g(Y)$ and $f(X)$ by $g(\frac{Y}{\mu})$ and $\mu f(\frac{X}{\nu})$ respectively where $\mu, \nu \neq 0, \in k$. We shall choose now μ and ν so that the theorem is reduced to a trivial case.

Let $g = (g_1, \ldots, g_p)$ and let $g_j = \sum_{\beta > 0} b_{j,\beta} Y^\beta$ be a local expansion of g_j about 0. Choose s so that each g_j converges in $P(s)$. We can find a constant N so that for all j and all β, $|b_{j,\beta}| s^\beta \leq N$. We let μ be an element of k such that $|\mu| > \max_j(\frac{1}{s_j}(1+N))$. Then for all j and all β:

$$\left| b_{j,\beta} \frac{1}{\mu^{|\beta|}} \right| < |b_{j,\beta}| \min_j (s_j)^{|\beta|} \frac{1}{1+N} \leq |b_{j,\beta}| s^\beta \frac{1}{1+N} < 1 \, .$$

Hence $g(\frac{Y}{\mu})$ has coefficients in the ring A_v of the valuation v of k and in particular $g(\frac{Y}{\mu})$ converges in $P_0(1)$.

By applying the above argument to μf, we may find $\nu \in k$ so that the local expansions of the coordinate functions $\mu f_i(\frac{X}{\nu})$ of $\mu f(\frac{X}{\nu})$ have coefficients in A_v.

We are therefore reduced to the case when the local expansions of the coordinate functions of f and g have coefficients in A_v. But then the formal series of the composite has coefficients in A_v and therefore converges in $P_0(1)$.

We now make explicit some facts about differentiation and Taylor series implicit in our discussion of the theorem that a convergent power series defines an analytic function.

Definition. Let $\phi : U \to V$ be a function, where $U \subset k^m$ and $V \subset k^n$ are open, and let $x \in U$. A linear function $L : k^m \to k^n$ is called a *derivative* of ϕ at x if:

$$|\phi(x+h) - \phi(x) - Lh| = o(|h|) \, , \qquad |h| \to 0$$

or, equivalently, if:

$$\lim_{\substack{|h| \to 0 \\ h \neq 0}} \frac{|\phi(x+h) - \phi(x) - Lh|}{|h|} = 0 \, .$$

Remarks. 1. If ϕ has a derivative L at x, then L is unique and is denoted by $D\phi(x)$.

2. If $D\phi(x)$ is applied to the vector $(0, \ldots, 1, \ldots, 0)$ which has 1 in the i-th place and 0 in all other places, the vector obtained in k^n is denoted by $D_i\phi(x)$ and is called the i-th partial derivative of ϕ at x.

To study the differentiability of analytic functions, it suffices first of all to restrict oneself to analytic functions with values in k and then to further restrict oneself to functions given by convergent power series since differentiability is a local property. We shall let δ_i denote the vector with 1 in the i-th place and 0 elsewhere.

Theorem. *Let $f = \sum a_\alpha X^\alpha$ be a power series convergent in $P_0(r)$ for $r > 0$. Then f is differentiable at each $x \in P_0(r)$ and we have:*

$$Df(x) = \begin{Bmatrix} \Delta^{\delta_1} f(x) \\ \vdots \\ \Delta^{\delta_n} f(x) \end{Bmatrix}$$

Hence, the derivative of an analytic function exists and is analytic, so that by induction an analytic function is infinitely differentiable.

Proof. The theorem follows immediately from the explicit calculations of the lemma of page 70 which show that $f(x+h) - f(x) - Df(x)h$ is a power series which is convergent in $P_0(r)$ and whose terms of degree 0 and 1 vanish.

Remarks. Let $D^\alpha = D_1^{\alpha_1} \cdots D_n^{\alpha_n}$.

1. $\alpha! \Delta^\alpha = D^\alpha$.

2. The expansion $f(x+h) = \sum_\beta \Delta^\beta f(x) h^\beta$ is just the Taylor series in characteristic 0.

3. $\binom{\alpha+\beta}{\alpha} \Delta^{\alpha+\beta} = \Delta^\alpha \Delta^\beta$.

We are now in a position to state:

Inverse Function Theorem. *Let $f : U \to k^n$ be analytic where U is open in k^n and suppose $0 \in U$ and $f(0) = 0$. Then, if $Df(0) : k^n \to k^n$ is a linear isomorphism, f is a local analytic isomorphism.*

Proof. The theorem is well known for $k = \mathbf{R}$ or $\mathbf{C}$ so we may assume by Ostrowski's Theorem that k has an ultrametric absolute value. Let $f = (f_1, \ldots, f_n)$. By following f with $Df(0)^{-1}$ if necessary, we may assume that:

$$f_i(X) = X_i - \sum_{\alpha > 1} a_{i,\alpha} X^\alpha = X_i - \phi_i(X)$$

Then, replacing $f(X)$ by $\mu f(\frac{X}{\mu})$ where $|\mu|$ is sufficiently large, we may also assume that $a_{i,\alpha} \in A_v$ for all i and α where A_v is the ring of the valuation of k.

To invert f, we seek convergent power series $\Psi_i(T)$ so that $X_i = \Psi_i(T)$ solves the equations:

$$(*) \qquad T_i = X_i - \phi_i(X)\ , \qquad 1 \le i \le n$$

We shall solve the problem in two steps:

1. We shall show that $(*)$ has a unique formal solution Ψ_i and we shall describe the relation between the coefficients of Ψ_i and those of ϕ_i.

2. We shall use two methods to show that the formal power series solution we have obtained converges.

Set $\Psi_i = \sum_{\beta>0} b_{i,\beta} T^\beta$ and consider the equations:

$$(**) \qquad \Psi_i(T) = T_i + \phi_i(\Psi(T)) \,, \qquad 1 \le i \le n$$

We see immediately that $b_{i,\delta_j} = 1$ or 0 according as $i = j$ or $i \neq j$. More generally, for arbitrary β, we see that $b_{i,\beta}$ is a linear combination with positive integral coefficients of monomials in the coefficients of Ψ and ϕ of degree *strictly less than* β. Moreover, the positive integral coefficients in this expression are independent of the $\{\phi_i\}$ and the $\{\Psi_i\}$. Hence, by induction, we see that:

$$b_{i,\beta} = p^i_\beta(a_{j,\alpha})$$

where:

1. p^i_β is a polynomial with positive integral coefficients independent of the $\{\phi_i\}$.

2. p^i_β depends only on the $a_{j,\alpha}$'s for $|\alpha| < |\beta|$.

The first method we give to prove convergence depends on the fact that we have assumed k ultrametric. By construction, $b_{i,\beta} \in A_v$ for all i and β. Hence the $\{\Psi_i\}$ converge in $P_0(1)$.

The second method which we give, Cauchy's method of majorants, works in the cases $k = \mathbf{R}$ or $\mathbf{C}$ as well. Suppose that we can find real positive power series $\{\bar\phi_i\}$ such that if $\{\bar\Psi_i\}$ is the formal solution for the inversion problem for $\{\bar\phi_i\}$ then:

1. $\bar\phi_i = \sum_{\alpha>1} \bar a_{i,\alpha} X^\alpha$ and $\bar\Psi_i = \sum_{\beta>0} \bar b_{i,\beta} T^\beta$ converge for $1 \le i \le n$.

2. For all i and α, $|a_{i,\alpha}| \le \bar a_{i,\alpha}$.

We shall then show that:

3. For all i and β, $|b_{i,\beta}| \le \bar b_{i,\beta}$.

It will then follow from 1 and 3 that Ψ_i converges for $1 \le i \le n$. To obtain 3 from 2, we simply note that since the p^i_β have positive integer coefficients, we have

$$|b_{i,\beta}| = |p^i_\beta(a_{j,\alpha})| \le P(|a_{j,\alpha}|) \le P(\bar a_{j,\alpha}) = \bar b_{i,\beta} \,.$$

It therefore remains to construct functions ϕ_i with the required properties.

For the case $n = 1$ and any positive integer m, a positive constant times the following function $\bar\phi^m$ will satisfy the first part of 1 as well as 2:

$$\bar\phi^m = \sum_{i>1} (mX)^i \qquad (m > 0).$$

We may by renormalizing the problem assume the positive constant is 1. Then we may explicitly compute the inverse function $\bar{\Psi}^m$ of $\bar{\phi}^m$. Indeed, we must solve the equation:

$$T = X - \frac{(mX)^2}{1 - mX} .$$

Its solution is given by:

$$\bar{\Psi}^m(T) = \frac{(1 + mT) - \sqrt{(1 + mT)^2 - 4(m^2 + m)T}}{2(m^2 + m)} ;$$

$\bar{\psi}^m(T)$ does indeed converge in a neighborhood of 0.

To solve the case for general n, we take $\bar{\phi}_i = \sum_{j>1}(X_1 + \cdots + X_n)^j$. Then the explicit solution $\bar{\Psi}_i$ for $\bar{\phi}_i$ is:

$$\bar{\Psi}_i = \frac{1}{n}\sum_{j \neq i}(T_i - T_j) + \bar{\Psi}^n\left(\frac{\sum T_j}{n}\right) .$$

Indeed:

$$\sum \bar{\Psi}_i = n\bar{\Psi}^n\left(\frac{\sum T_j}{n}\right)$$

$$\bar{\Psi}_i - \bar{\phi}_i(\bar{\Psi}) = \frac{1}{n}\sum_{j \neq i}(T_i - T_j) + \bar{\Psi}^n\left(\frac{\sum T_j}{n}\right) - \bar{\phi}^n\left(\bar{\Psi}^n\left(\frac{\sum T_j}{n}\right)\right)$$

$$= \frac{1}{n}\sum_{j \neq i}(T_i - T_j) + \frac{1}{n}\left(\sum T_j\right) = T_i .$$

Since the $\bar{\Psi}_i$ converge in a neighborhood of 0 the theorem is proved.

"Tournants dangereux"

1. Suppose k is ultrametric. Then the function ϕ which is 1 on A_v and 0 outside A_v is everywhere analytic. This follows from the fact that A_v is both open and closed.
2. If k has characteristic $p > 0$, then for an analytic function ϕ defined on k^n and for $|\alpha| \geq (p-1)n + 1$, one has $D^\alpha\phi = 0$. In particular, the radius of convergence of the derivative of a function may be strictly greater than that of the function.
3. If ϕ is analytic on $U \subset k^n$, $x \in U$, and $P_0(r)(x) \subset U$, then the local expansion of ϕ at x does not necessarily converge on all of $P_0(r)$. In general, this is only true for $k = \mathbf{C}$.

Chapter III. Analytic Manifolds

We denote by k a field complete with respect to a non-trivial absolute value.

1. Charts and atlases

Let X be a topological space.

A *chart* c on X is a triple $c = (U, \phi, n)$ such that:

1. $U \subset X$ is open
2. $n \in \mathbf{Z}$ and $n \geq 0$.
3. $\phi : U \to \phi U \subset k^n$ is open and ϕ is a homeomorphism.

Notation.
$U = O(c) =$ Open set of c.
$\phi = M(c) =$ Map of c.
$n = \dim_k c =$ Dimension of c.

Let $c = (U, \phi, n)$ and $c' = (U', \phi', n')$ be charts on X. Then c and c' are said to be *compatible* if, setting $V = U \cap U'$, the maps $\phi' \circ \phi^{-1}|_{\phi(V)}$ and $\phi \circ \phi'^{-1}|_{\phi'(V)}$ are analytic.

$$\begin{array}{ccc} & & \phi(V) \subset k^n \\ & \overset{\phi}{\nearrow} & \\ V & & \phi'\circ\phi^{-1}\downarrow\uparrow\phi\circ\phi'^{-1} \\ & \underset{\phi'}{\searrow} & \\ & & \phi'(V) \subset k^{n'} \end{array}$$

If c and c' are compatible, then $V \neq \emptyset$ implies $n = n'$.

A family $\{c_i\}_{i \in I}$ of charts on X is said to *cover* X if $\bigcup_{i \in I} O(c_i) = X$.

An *atlas* A on X is a family of charts on X which covers X and such that the charts in the family are mutally compatible.

Two atlases A and A' on X are said to be *compatible* if equivalently:

1. $A \cup A'$ is an atlas.
2. If $c \in A$ and $c' \in A'$, then c and c' are compatible.

Remark. Compatibility of atlases is an equivalence relation. Indeed, the reflexive and symmetric properties of an equivalence are obvious. To check transitivity, let A_1, A_2, and A_3 be atlases and let $c_1 \in A_1$ and $c_3 \in A_3$. We must show that c_1 and c_3 are compatible. Let $V = O(c_1) \cap O(c_3)$. If $V = \emptyset$, then c_1 and c_3 are trivially compatible. Suppose $V \neq \emptyset$ and let $\phi_1 = M(c_1)$ and $\phi_3 = M(c_3)$. It suffices, by symmetry, to check that $\phi_3 \circ \phi^{-1}$ is analytic on $\phi_1(V)$. We shall check that this map is analytic at $\phi_1(x)$ for each $x \in V$. Choose $c_2 = (U, \phi, n) \in A_2$ such that $x \in U$. Then:

$$\phi \circ \phi_1^{-1} : \phi_1(U \cap V) \to \phi(U \cap V) \text{ is analytic at } \phi_1(x).$$
$$\phi_3 \circ \phi^{-1} : \phi(U \cap V) \to \phi_3(U \cap V) \text{ is analytic at } \phi(x).$$

Hence $\phi_3 \circ \phi_1^{-1} = (\phi_3 \circ \phi^{-1}) \circ (\phi \circ \phi_1^{-1})$ is analytic at $\phi_1(x)$ as desired.

2. Definition of analytic manifolds

Let X be a topological space.

An *analytic manifold* structure on X is an equivalence class of compatible atlases on X.

An alternate definition may be given as follows. Say that an atlas A on X is *full* if whenever c is a chart on X such that c is compatible with all charts $c' \in A$ then $c \in A$. Then it is clear that each equivalence class of atlases on X contains exactly one full atlas. We may therefore define:

An *analytic manifold* structure on X is the choice of a full atlas on X.

Henceforth, in this chapter, X will denote a topological space with a fixed analytic manifold structure. $A(X)$ will denote its full atlas. A chart c on X will mean a chart belonging to $A(X)$.

If X is an analytic manifold, and $x \in X$, $\dim_x X$ is defined as the dimension of any chart c on X such that $x \in O(c)$; it is called the *dimension of X at x*. The function $x \mapsto \dim_x X$ is locally constant on X; if it is constant, and equal to n, one says that x is everywhere of dimension n.

It is customary to introduce special terminology in the cases which are of greatest interest:

When $k = \mathbf{R}$, we say that X is a "real analytic" manifold.

When $k = \mathbf{C}$, we say that X is a "complex analytic" manifold.

When $k = \mathbf{Q}_p$, p prime in $\mathbf{Z}$, we say that X is a "p-adic analytic" manifold.

3. Topological properties of manifolds

Let $x \in k^n$, $n \in \mathbf{Z}$ and $n \geq 0$, and let $r \in \mathbf{R}$, $r > 0$. Then the ball, $B(r)(x)$, of radius r about x is the polydisk $P(s)(x)$ where $s = (r, \ldots, r)$.

Let B be a subset of X. Then B is said to be a *ball* if there is a chart $c = (U, \phi, n)$ such that $B \subset U$ and ϕB is a ball in k^n. Then:

1) Every point $x \in X$ has a neighborhood B which is a ball. In particular, X is locally a complete metric space (hence a Baire space).

2) Suppose k locally compact. Then a ball in X is compact. In particular, if X is Hausdorff, then X is locally compact.

3) Suppose X is regular and k is ultrametric. Then each $x \in X$ has a basis of neighborhoods which are both open and closed.

The only property which is perhaps not immediately obvious is 3. To prove 3, let B be a ball containing x and let $c = (U, \phi, n)$ be a chart such that $B \subset U$ and ϕB is a ball in k^n. Now ϕB is open in k^n so that B is open in X. Hence, since X is regular, there is a neighborhood V of x such that $V \subset B$ and V is closed in X. Then the inverse image under ϕ of balls in ϕV containing ϕx is a fundamental system of neighborhoods of x which are both open and closed.

Remark. In Appendix 1 to this chapter, an example due to George Bergman is given in which the conclusion of 3 fails when X is only Hausdorff.

4. Elementary examples of manifolds

1) $X =$ discrete space ($n = 0$).

2) $X = V$, where V is a finite dimensional vector space over k, $\dim_k V = n$. Let A be the collection of charts $c = (V, \phi, n)$ on V where $\phi : V \to k^n$ is a linear isomorphism. Then the charts in A are compatible so that A is an atlas. We give V the manifold structure determined by A.

3) Let X be a manifold and let U be open in X. Let $A = A(X)$. Define:

$$A_U = \left\{ c \in A : O(c) \subset U \right\}.$$

Then A_U defines a full atlas on U. The space U together with this atlas is called an open submanifold of X.

4) Let X be a topological space and let $X = \bigcup_{i \in I} U_i$. Suppose:

a. Each U_i is open in X.

b. On each U_i, there is given a structure of analytic manifold.

c. For each i and j, the manifold structures on $U_i \cap U_j$ induced by the manifold structures on U_i and U_j agree.

Then, on X there is a unique manifold structure such that its restriction to each U_i is the given one.

5) The line with a "point doubled". Let $k = \mathbf{R}$. Then we define a manifold X by identifying two copies of $\mathbf{R}$ at all points except 0. We shall interpret this space as a quotient space.

First consider the plane, $\mathbf{R}^2$, as being fibred by lines. Then the quotient space obtained by collapsing the fibres is $\mathbf{R}$.

Now suppose we remove the origin from $\mathbf{R}^2$ and collapse the connected components of the fibres. Then we get precisely the line with 0 "doubled".

Note that the manifold in this example is not Hausdorff.

5. Morphisms

Let X and Y be two analytic manifolds. A function $f : X \to Y$ is said to be an *analytic function* or *morphism* if:

1. f is continuous.

2. f is "locally given by analytic functions", that is, there exists atlases A of X and B of Y such that if $c = (U, \phi, m) \in A$ and $d = (V, \psi, n) \in B$, then, setting $W = U \cap f^{-1}V$, the composite

$$\phi W \xrightarrow{\phi^{-1}} V \xrightarrow{f} V \xrightarrow{\psi} \psi V$$

is analytic.

Remarks. 1) We describe condition 2 by saying that f is "locally given by analytic functions" since, in coordinates, composites of the form $\psi \circ f \circ \phi^{-1}$ may be written as n-tuples of analytic functions of m variables.

2) Condition 2 is independent of the choice of atlases A and B, as is seen by an argument similar to the one showing that compatibility of atlases is an equivalence relation.

The following formal properties of morphisms are easily verified:

1) The composition of morphisms is a morphism.

2) The identity map on a manifold is a morphism.

3) Suppose $f : X \to Y$ and $g : Y \to X$ are maps such that $g \circ f = 1_X$ and $f \circ g = 1_Y$. Then f is an isomorphism if and only if f and g are morphisms.

Let us quote without proof a much deeper statement:

Theorem. *Assume k is algebraically closed of characteristic zero, and let $f : X \to Y$ be a morphism of analytic manifolds. If f is an homeomorphism, then f is an analytic isomorphism.*

Remark. The conclusion of the theorem is false for $k = \mathbf{R}$, as the example $f : \mathbf{R} \to \mathbf{R}$ given by $f(x) = x^3$ shows.

6. Products and sums

1) *Products*

Let $\{X_i\}_{i\in I}$ be a *finite* collection of manifolds and let A_i be an atlas for X_i, for each $i \in I$. Suppose $c_i \in A_i$, for each $i \in I$, $c_i = (U_i, \phi_i, n_i)$. Define $\prod_{i\in I} c_i$ by:

$$\prod_{i\in I} c_i = \left(\prod_{i\in I} U_i, \prod_{i\in I} \phi_i, \sum_{i\in I} n_i\right) .$$

Set:

$$X = \prod_{i\in I} X_i \,, \qquad A = \Big\{ \prod_{i\in I} c_i : c_i \in A_i, \ i \in I \Big\} .$$

Then X is a topological space and A is an atlas on X. The space X together with the manifold structure determined by A is called the product of $\{X_i\}_{i\in I}$. The usual universal property for products, namely, for all manifolds Y,

$$\operatorname{Mor}\Big(Y, \prod_{i\in I} X_i\Big) = \prod_{i\in I} \operatorname{Mor}(Y, X_i)$$

is easily verified.

2) *Sums or disjoint unions.*

Let $\{X_i\}_{i\in I}$ be an arbitrary family of manifolds. Let $\sum_{i\in I} X_i$ or $\coprod_{i\in I} X_i$ denote the disjoint union of the topological spaces X_i. Then, by Example 4 of 3.4, there is an unique manifold structure on $X = \sum_{i\in I} X_i$ compatible with the manifold structure on each X_i and we furnish X with this manifold structure. X is then called the *sum* or *disjoint union* of $\{X_i\}_{i\in I}$. The usual property for sums, namely, for all manifolds Y,

$$\operatorname{Mor}\Big(\coprod_{i\in I} X_i, Y_i\Big) = \prod_{i\in I} \operatorname{Mor}(X_i, Y)$$

is easily verified.

In Appendix 2 to this chapter, the structure of compact manifolds defined over locally compact, ultrametric fields k is described in detail using the notion of disjoint union of manifolds.

7. Germs of analytic functions

Let $x \in X$ and let $\underline{F}_x$ be the set of pairs (U, ϕ) where U is an open neighborhood of x and ϕ is an analytic function on U. The set $\underline{F}_x$ is called the set of *local functions* at x. We introduce an equivalence relation on $\underline{F}_x$ as follows:

We say that two elements (U, ϕ) and (V, ψ) of $\underline{F}_x$ are equivalent if there is an open neighborhood W of x such that $W \subset U \cap V$ and $\phi|_W = \psi|_W$. The set of equivalence classes of $\underline{F}_x$ is denoted $\underline{H}_x$ and is called either the set of *germs of analytic functions* at x or the *local ring* at x.

We define addition and multiplication in $\underline{H}_x$ as follows. Let f and g be germs of functions at x. Choose $(U, \phi) \in f$ and $(V, \psi) \in g$. Let $W = U \cap V$. Then $f + g$ is defined to be the class of $(W, f|_W + g|_W)$ while $f \cdot g$ is defined to be the class of $(W, (f|_W) \cdot (g|_W))$. It is easily seen that these definitions are independent of the choices made.

There is a canonical map $k \to \underline{F}_x$ which sends $\alpha \in k$ to (X, c_α) where c_α is the constant function α on X. This map induces a canonical inclusion $\imath : k \to \underline{H}_x$ which makes $\underline{H}_x$ a k-algebra.

There is also a canonical map $\underline{F}_x \to k$ which sends $(U, \phi) \in \underline{F}_x$ to $\phi(x)$. This map induces a canonical homomorphism $\theta : \underline{H}_x \to k$ of $\underline{H}_x$ onto k. For $f \in \underline{H}_x$, we let $f(x)$ denote $\theta(f)$ and we call $f(x)$ the value of f at x. The kernel $\mathfrak{m}_x$ of θ is a maximal ideal.

Since $\theta \circ \imath = \mathrm{Id}_k$, there is a canonical decomposition $\underline{H}_x = \imath(k) \oplus \mathfrak{m}_x$. We shall identify k with $\imath(k)$ and suppress the mention of $\imath$.

We shall now show that $\underline{H}_x$ is a local ring by means of the following stronger statement:

Lemma. *Let (U, ϕ, n) be a chart at x. Then ϕ induces, via composition of functions, an isomorphism $\bar{\phi} : \underline{H}_0 \to \underline{H}_x$ such that $\bar{\phi}(\mathfrak{m}_0) = \mathfrak{m}_x$. Here $\underline{H}_0$ is the ring of germs of functions at 0 in k^n and $\mathfrak{m}_0$ is its maximal ideal. $\underline{H}_0$ is isomorphic to the local ring of convergent power series in n variables.*

Proof. All statements are clear except perhaps the statement that the ring of convergent power series in n variables is local. To prove this, we must show that if f is a convergent power series such that $f(0) \neq 0$ then f is invertible.

We may assume that $f = 1+\psi$ where $\psi(0) = 0$. Then, since the map $g(x) = \frac{1}{x}$ is analytic near 1, $\frac{1}{f} = g \circ f$ is analytic near 0.

Suppose $f \in \underline{H}_x$ and $f \neq 0$. Then we define $\mathrm{ord}_x f$ to be the least integer μ such that $f \notin \mathfrak{m}_x^{\mu+1}$. The preceding lemma shows that for any chart (U, ϕ, n) $\mathrm{ord}_x f$ is the least integer μ such that $\bar{\phi}(f)$ has in its power series expansion a non-vanishing homogeneous term of total degree μ.

8. Tangent and cotangent spaces

Let $x \in X$. Define:

$$T_x^* X = \mathfrak{m}_x/\mathfrak{m}_x^2 = \text{cotangent space of } X \text{ at } x.$$
$$T_x X = (\mathfrak{m}_x/\mathfrak{m}_x^2)^* = \text{tangent space of } X \text{ at } x.$$

We give two alternate descriptions of $T_x X$:

1) $T_x X$ is canonically isomorphic to the space of derivations $v : \underline{H}_x \to k$.

Let $v \in T_x X$. Then v defines a linear form on $\mathfrak{m}_x$ which vanishes on $\mathfrak{m}_x^2$. Extend v to a linear form on $\underline{H}_x = k \oplus \mathfrak{m}_x$ by setting $v = 0$ on k. Then v is a derivation on $\underline{H}_x$. Indeed, v is k-linear. Hence it remains to check that for $f, g \in \underline{H}_x$:

$$v(fg) = (vf)g(x) + f(x)(vg) .$$

Since both sides of the equation are bilinear, it suffices to check the equation in three special cases:

1. $f, g \in k$.
2. $f \in k$ and $g \in \mathfrak{m}_k$ or $f \in \mathfrak{m}_x$ and $g \in k$.
3. $f, g \in \mathfrak{m}_x$.

Cases 1 and 3 follow since both sides of the equation are 0. Case 2 is a consequence of the linearity of v and the fact that v vanishes on k.

Conversely, given a derivation d of $\underline{H}_x$, d vanishes on k and $\mathfrak{m}_x^2$. Hence d comes from a unique linear form in $\mathfrak{m}_x/\mathfrak{m}_x^2$, that is, from a unique tangent vector v. This establishes the desired isomorphism.

2) $T_x X$ is canonically isomorphic to the space $\underline{C}_x$ of "tangency classes of curves at x".

We first define $\underline{C}_x$ precisely. Let $\underline{F}'_x$ be the set of pairs (N, ψ) where N is an open neighborhood of 0 in k and $\psi : N \to X$ is such that $\psi(0) = x$. We define an equivalence relation in $\underline{F}'_x$ as follows. Let $(N_i, \psi_i) \in \underline{F}'_x$, $i = 1, 2$. Choose a chart (U, ϕ, n) at x. Then, for $i = 1, 2$, $\phi \circ \psi$ is defined in the neighborhood $N_i \cap \psi_i^{-1}(U)$ of 0. We say that (N_1, ψ_1) is equivalent to (N_2, ψ_2) if $D(\phi \circ \psi_1)(0) = D(\phi \circ \psi_2)(0)$. We let $\underline{C}_x$ denote the set of equivalence classes of elements of $\underline{F}'_x$.

Notice that the map which sends $(N, \psi) \in \underline{F}'_x$ to $D(\phi \circ \psi)(0)$ induces a bijection $\bar{\phi} : \underline{C}_x \to L(k, k^n) = k^n$. Hence $\underline{C}_x$ may be given the structure of a vector space over k.

It is easily verified that the definition of $\underline{C}_k$ and the definition of the vector space structure on $\underline{C}_x$ are independent of the choice of (U, ϕ, n). Indeed, suppose (U', ϕ', n) is a second chart at x. Then, for $(N, \psi) \in \underline{F}'_x$, $D(\phi' \circ \psi)(0) = D(\phi' \circ \phi^{-1})(0) \circ D(\phi \circ \psi)(0)$. It is clear from this formula that the equivalence defining $\underline{C}_x$ is independent of the choice of the chart. From this formula, it also follows that $\bar{\phi}' = D(\phi' \circ \phi^{-1})(0) \circ \bar{\phi}$, so that the vector space structure on $\underline{C}_x$ is well defined.

We now define a bilinear pairing $\underline{C}_x \times T^*_x X \xrightarrow{\omega} k$ by defining first a pairing $\underline{F}'_x \times \underline{F}_x \to k$. The latter pairing sends $(N, \psi) \in \underline{F}'_x$ and $(V, f) \in \underline{F}_x$ to $D(f \circ \psi)(0) \in k$. This pairing clearly induces a pairing $\underline{C}_x \times T^*_x X \xrightarrow{\omega} k$, which is bilinear and establishes $\underline{C}_x$ as the dual of $T^*_x X$.

Remarks. (1) Intuitively the pairing ω is simply differentiation of a function in the direction of the tangent to a curve.

(2) The process of defining a linear space structure on $\underline{C}_x$ would fail if we wished to construct a space of higher derivatives to curves. The reason is that the higher derivatives of the composite of two functions are not bilinear in the derivatives of each of the functions.

Example. X is a finite dimensional vector space V. Then:

$$T_x V = L(k, V) = V$$
$$T^*_x V = L(V, k) = V^* .$$

We shall now define the related concepts of differentials of a function and tangent map to a morphism.

Let $f \in \underline{H}_x$. Then $f - f(x) \in \mathbf{m}_x$. The image of $f - f(x)$ in $\mathbf{m}_x / \mathbf{m}_x^2 = T^*_x X$ is called the *differential* of f at x and is denoted by df_x. Let $v \in T_x X$. Then v applied to df_x is called the derivative of f in the direction v and is denoted by $\langle v, df_x \rangle$ or $v \cdot f_x$; we may think of df_x as a linear form on $T_x X$.

Let f be a function defined in a neighborhood of x. Then f defines an element of $\underline{H}_x$ and hence a linear form df_x on $T_x X$.

Let Y be a second manifold, let $y \in Y$, and let $\phi : X \to Y$ be a morphism such that $\phi(x) = y$. Define $T_x \phi : T_x X \to T_y Y$ by the formula:

$$\langle T_x \phi(v), df_y \rangle = \langle v, d(f \circ \phi)_x \rangle ,$$

for all $v \in T_x X$ and all $f \in \underline{H}_x$. Equivalently, we can define $T_x \phi$ by defining its transpose $T^*_x \phi : T^*_y Y \to T^*_x X$. For $f \in \underline{H}_x$, we define $T^*_x \phi(df_y) = d(f \circ \phi)_x$. The linear map $T_x \phi$ is called the *tangent map* of ϕ.

In the special case when $Y = k$ and ϕ is a function f, then $T_x f = df_x$.

We conclude this section by examining tangent spaces of products. Let X, Y, and Z be manifolds and let $x \in X$, $y \in Y$, and $z \in Z$. Then:

$$T_{x,y}(X \times Y) = T_x X \times T_y Y$$
$$T^*_{x,y}(X \times Y) = T^*_x X \times T^*_y Y .$$

Let $\phi : X \times Y \to Z$ be a morphism such that $\phi(x,y) = z$. Then $T_{x,y}\phi$ defines $T^X_{x,y}\phi : T_xX \to T_zZ$ and $T^Y_{x,y}\phi : T_yY \to T_zZ$ by the conditions:

$$T_{x,y}\phi(v,w) = T^X_{x,y}\phi(v) + T^Y_{x,y}\phi(w) .$$

The maps $T^X\phi$ and $T^Y\phi$ are called the partial derivatives of ϕ along X and Y respectively.

9. Inverse function theorem

Let $x \in X$ and let $f_1, \dots, f_m$ be analytic functions on a neighborhood U of x. Let $F(y) = (f_1(y), \dots, f_m(y))$ for $y \in U$. We say that $\{f_i\}_{1 \le i \le m}$ defines a coordinate system at x if there exists an open neighborhood U' of x, contained in U, such that $(U', F|_{U'}, m)$ is a chart on X.

Theorem 1. *The following are equivalent:*

1. $\{f_i\}$ *defines a coordinate system at* x.
2. df_{ix} *form a basis of* T^*_xX.

Theorem 1 is a consequence of the following more general theorem:

Theorem 2. *Let* X *and* Y *be manifolds,* $x \in X$ *and* $y \in Y$*, and let* $\phi : X \to Y$ *be a morphism such that* $\phi(x) = y$*. Then the following are equivalent:*

1. ϕ *is a local isomorphism at* x.
2. $T_x\phi$ *is an isomorphism.*

2′. $T^*_x\phi$ *is an isomorphism.*

Proof. $1 \Rightarrow 2$ and $2 \Rightarrow 2'$ are obvious.

$2 \Rightarrow 1$: This is a local question and the result has been proved in the local case in Chapter 2.

Definition. A morphism ϕ satisfying the equivalent conditions of Theorem 2 at x is said to be *étale at* x. If ϕ is étale at x for all $x \in X$, ϕ is said to be *étale.*

10. Immersions, submersions, and subimmersions

Let X and Y be manifolds, $x \in X$ and $y \in Y$, and let $\phi : X \to Y$ be a morphism such that $\phi(x) = y$. Let $m = \dim_k x$ and $n = \dim_k y$.

Definition. Let $\bar{X}$ and $\bar{Y}$ be manifolds, $\bar{x} \in \bar{X}$ and $\bar{y} \in \bar{Y}$, and let $\bar{\phi} : \bar{X} \to \bar{Y}$ be a morphism such that $\bar{\phi}(\bar{x}) = \bar{y}$. Then (X, Y, x, y, ϕ) *looks locally like* $(\bar{X}, \bar{Y}, \bar{x}, \bar{y}, \bar{\phi})$ if there exist open neighborhoods U of x, V of y, $\bar{U}$ of $\bar{x}$, $\bar{V}$ of $\bar{y}$ and isomorphisms $g : U \to \bar{U}$ and $h : V \to \bar{V}$ such that:

1. $\phi U \subset V$ and $\bar\phi \bar U \subset \bar V$.

2. $g(x) = \bar x$ and $h(y) = \bar y$.

3. The following diagram is commutative:

$$\begin{array}{ccc} U & \xrightarrow{\phi} & V \\ {\scriptstyle g}\downarrow & & \downarrow{\scriptstyle h} \\ \bar U & \xrightarrow{\bar\phi} & \bar V \end{array} .$$

Remark. We shall apply this definition mainly when $\bar X$ is a linear space E, $\bar Y$ is a linear space F, and $\bar\phi$ is a linear map. In this case, we will take $\bar x = 0$ and $\bar y = 0$ without explicit mention.

1) Immersions

Theorem. *The following are equivalent:*

1. *$T_x\phi$ is injective.*

2. *There exist open neighborhoods U of x, V of y, and W of 0 (in k^{n-m}) and an isomorphism $\psi : V \to U \times W$ such that:*
   a. *$\phi U \subset V$*
   b. *If $\imath$ denotes the inclusion $U \to U \times \{0\} \subset U \times W$, then the following diagram is commutative:*

$$\begin{array}{ccc} U & \xrightarrow{\phi} & V \\ & {\scriptstyle \imath}\searrow & \downarrow{\scriptstyle \psi} \\ & & U \times W \end{array} .$$

3. *(X, Y, x, y, ϕ) looks locally like a linear injection $\bar\phi : E \to F$ where E and F are m and n dimensional vector spaces respectively.*

4. *There exist coordinates $\{f_i\}$ at x and $\{g_j\}$ at y such that $f_i = g_i \circ \phi$ for $1 \le i \le m$ and $0 = g_j \circ \phi$ for $m + 1 \le j \le n$.*

5. *There exist open neighborhoods U of x and V of y, and a morphism $\sigma : V \to U$ such that $\phi U \subset V$ and $\sigma \circ \phi = \mathrm{id}_U$.*

Proof. The implications $2 \Rightarrow 3 \Rightarrow 4 \Rightarrow 5 \Rightarrow 1$ are elementary.

We show $1 \Rightarrow 2$. Since the question is local, we may assume that the following conditions are satisfied:

a. Y is an open subset of k^n.

b. $\phi(x) = 0$ and $\operatorname{Im} T_x\phi = k^m \times \{0\} \subset k^m \times k^{n-m} = k^n$.

Let W be $\{0\} \times k^{n-m} \subset k^n$. Define $\phi' : X \times W \to Y$ by $\phi'(x, w) = \phi(x) + w$. Then by the inverse function theorem, ϕ' is a local isomorphism at x. Hence, by shrinking X, Y and W, we may assume that ϕ' is an isomorphism. The inverse ψ of ϕ' satisfies the conditions of 2.

Definition. A morphism ϕ satisfying the equivalent conditions of the preceding theorem at x is called an *immersion at* x. A morphism ϕ which is an immersion at all $x \in X$ is called an *immersion.*

2) Submersions

Theorem. *The following are equivalent:*

1. *$T_x\phi$ is surjective.*

2. *There exist open neighborhoods U of x, V of y and W of 0 in k^{m-n}) and an isomorphism $\psi : U \to V \times W$ such that*
   a. *$\phi U = V$.*
   b. *If p denotes the projection $V \times W \to V$, then the following diagram is commutative:*

$$\begin{array}{ccc} U & \xrightarrow{\phi} & V \\ \psi\downarrow & \nearrow p & \\ V \times W & & \end{array}$$

3. *(X, Y, x, y, ϕ) looks locally like a linear surjection $\bar{\phi} : E \to F$ where E and F are m and n dimensional vector spaces respectively.*

4. *There exist coordinates $\{f_i\}$ at x and $\{g_j\}$ at y such that $f_i = g_i \circ \phi$ for $1 \leq i \leq n$.*

5. *There exist open neighborhoods U of x and V of y and a morphism $\sigma : V \to U$ such that $\phi U \subset V$ and $\phi \circ \sigma = \mathrm{id}_V$.*

Proof. The proof is similar to the proof of the corresponding theorem on immersions and is left as an exercise to the reader.

Definition. A morphism ϕ satisfying the equivalent conditions of the preceding theorem at x is called a *submersion at* x. A morphism ϕ which is a submersion at all $x \in X$ is called a *submersion.*

3) Remarks. 1. Étale is equivalent to immersion and submersion.

2. The use of the word "immersion" is relatively common (Whitney, Smale). "Submersion" is a Bourbaki innovation, reproduced in Lang's book. Sometimes the phrase "ϕ has maximal rank" (meaning $T_x\phi$ is injective if $m \leq n$ and $T_x\phi$ is surjective if $m \geq n$) is used to include both concepts.

3. An *embedding* is a morphism ϕ such that:
   a. ϕ is an immersion.
   b. $X \to \phi(X)$ is a homeomorphism.

4) Subimmersions

Definition. ϕ is a *subimmersion at* x if the following equivalent conditions are satisfied:

1. ϕ looks locally like a composition $X \xrightarrow{s} Z \xrightarrow{\imath} Y$ where s is a submersion and $\imath$ is an immersion.

2. ϕ looks locally like a linear map $\bar{\phi} : E \to F$ where E and F are vector spaces of dimension m and n respectively.

A morphism ϕ which is a subimmersion at all $x \in X$ is called a *subimmersion.*

Remarks. 1. The set of points $x \in X$ where a morphism $\phi : X \to Y$ is an immersion (resp. a submersion, a subimmersion) is *open* in X.

2. The composition of two immersions (resp. submersions) is an immersion (resp. a submersion). The analogous statement for subimmersions is false.

Theorem. *Assume* $\operatorname{char} k = 0$. *Then the following are equivalent:*

1. *ϕ is a subimmersion at x.*

2. $\operatorname{rank} T_{x'}\phi$ *is constant for $x' \in U$ and U some neighborhood of x.*

Proof. $1 \Rightarrow 2$: Clear.

$2 \Rightarrow 1$: Let $p = \dim_k \operatorname{Im} T_x\phi$. Then, since the question is local, we may assume that the following conditions are satisfied:

a. $Y = V_1 \times V_2$ is open in $k^p \times k^{n-p}$.

b. $\phi(x) = 0$ and $\operatorname{Im} T_x\phi = k^p \times \{0\}$.

Let $\pi : k^p \times k^{n-p} \to k^p$ be the projection on the first factor. Then $\pi \circ \phi$ is a submersion. Hence we may further assume that:

c. $X = V_1 \times U_2$ is open in $k^p \times k^{m-p}$.

d. $\pi \circ \phi : V_1 \times U_2 \to V_1$ is the projection on the first factor.

The morphism ϕ then has the following form:

$$\phi(x_1, x_2) = (x_1, \psi(x_1, x_2))$$

Finally since $T_{x'}\phi$ has locally constant rank we may assume that the rank of $T_{x'}\phi$ is in fact constant on $V_1 \times U_2$ (rank $= p$).

We contend that ψ must be independent of x_2 in a neighborhood of zero. Indeed, $D_2\psi(x_1, x_2) = 0$ since otherwise ϕ would have rank greater than p at (x_1, x_2). Our contention is therefore a consequence of the following lemma:

Lemma. *Let $f : U \times V \to k$ be a function such that $D_2 f$ is identically* 0. *Then,* $\operatorname{char} k = 0$ *implies that f is locally independent of the V coordinate.*

Proof. Write f locally as a power series $\sum f_\alpha(y)x^\alpha$. Then $D_2 f = 0$ implies $D_2 f_\alpha = 0$ for all α. We must show $f_\alpha = c_\alpha$ where c_α is a constant. We have therefore reduced the theorem to the case $f = f_\alpha$. Write $f = \sum b_\beta y^\beta$. Then $Df = 0$ implies $\beta_i b_\beta = 0$ where $\beta = (\beta_1, \ldots, \beta_r)$ and $1 \le i \le r$. Hence, since $\operatorname{char} k = 0$, $b_\beta = 0$ for $\beta \neq 0$. Hence f is constant.

We conclude the proof of the theorem by noting that ϕ may now be written as $V_1 \times U_2 \to V_1 \to V_1 \times V_2$ where the first map is pr_1 and the second is $\mathrm{Id}_{V_1} \times \psi$. The first map is a submersion and the second is an immersion.

Corollary 1. *Assume* $\operatorname{char} k = 0$. *The set of points $x \in X$ where ϕ is a subimmersion is dense in X.*

Let X' be this set, and put $f(x) = \operatorname{rank} T_x\phi$. By the previous theorem, X' is the set of elements of X where f is locally constant. The fact that X' is dense follows then immediately from the following two properties of f:

a) f takes integral values, and is locally bounded.

b) f is lower semi-continuous.

Corollary 2. *Assume* $\operatorname{char} k = 0$ *and ϕ is injective. The set of points $x \in X$ where ϕ is an immersion is dense in X.*

This follows from Cor. 1 and the fact that an injective subimmersion is an immersion.

11. Construction of manifolds: inverse images

1) A uniqueness principle

Theorem. *Let X be a topological space, let A and B be full atlases on X, and let X_A (resp. X_B) denote the manifold whose underlying space is X that is determined by A (resp. B). Then the following are equivalent:*

1. $X_A = X_B$, *that is,* $A = B$.

2. *For all manifolds Y,* $\mathrm{Mor}(X_A, Y) = \mathrm{Mor}(X_B, Y)$.

3. *For all manifolds Y,* $\mathrm{Mor}(Y, X_A) = \mathrm{Mor}(Y, X_B)$.

Proof. The theorem is a special case of the theorem which states that an object which represents a functor is determined up to a unique isomorphism. Nevertheless, we give the proof in this case.

$1 \Rightarrow 2$: Trivial.

$2 \Rightarrow 1$: Setting $Y = X_A$, we see that $\mathrm{Id}_X : X_B \to X_A$ is a morphism. Similarly, $\mathrm{Id}_X : X_A \to X_B$ is a morphism. A and B are therefore compatible atlases and hence $A = B$ since A and B are full.

The proof of $1 \iff 3$ is equally simple.

We now state two lemmas which we will use in the application of the preceding theorem. We let X and Y be manifolds and $f : X \to Y$ be a morphism.

Lemma 1. *Suppose f is an immersion. Then:*

$$g \in \mathrm{Mor}(Z,X) \iff \begin{array}{l} \text{a. } g \textit{ is continuous} \\ \text{b. } f \circ g \in \mathrm{Mor}(Z,Y) \end{array}.$$

Lemma 2. *Suppose f is a submersion. Then:*

1. *f is an open map. In particular, $f(X)$ is open in Y.*
2. *Suppose that $f(X) = Y$. Then*

$$g \in \mathrm{Mor}(Y,Z) \iff g \circ f \in \mathrm{Mor}(X,Z)\ .$$

Lemmas 1 and 2 are an immediate consequence of the local description of immersions and submersions which we have given in III.10.

2) Inverse image constructions

Let X be a topological space, Y be a manifold, and $f : X \to Y$ be a continuous map.

Theorem 1. *If there exists a manifold structure on X such that f is an immersion then this manifold structure is unique.*

Proof. By Lemma 1 of n°1, for all Z, $\mathrm{Mor}(Z,X)$ is determined by the topological structure of X and the manifold structure of Y. Hence, by the Theorem of n°1, the manifold structure on X is unique.

Let $x \in X$. We say that (X, f) satisfies (Im) at x if the following condition is satisfied:

(Im): *There exists an open neighborhood U of x in X, a chart $c = (V, \phi, n)$ of Y, and a linear subspace E of k^n such that:*

a. *$f(U) \subset V$ and f is a homeomorphism of U onto $f(U)$.*
b. *$\phi(f(U)) = E \cap \phi(V)$.*

Theorem 2. *The following are equivalent:*

1. *There exists a manifold structure on X such that f is an immersion.*
2. *The pair (X, f) satisfies* (Im).

Proof. $1 \Rightarrow 2$: Part 3 of the Theorem of III.10, n° 1.

$2 \Rightarrow 1$: Choose an open covering $\{U_i\}_{i \in I}$ of X such that, for each $i \in I$, there exists a chart $c_i = (V_i, \phi_i, n_i)$ and a linear subspace E_i of k^{n_i} satisfying:

a. $f(U_i) \subset V$ and f is a homeomorphism of U_i onto $f(U_i)$.

b. $\phi_i(f(U_i)) = E_i \cap \phi_i(V_i)$.

Then there exists a manifold structure on U_i such that $f|_{U_i}$ is an immersion. Moreover, on $U_i \cap U_j$, the manifold structures induced from U_i and U_j agree, by Theorem 1. Hence, by III.4, n° 4, there is a manifold structure on X compatible with the manifold structure on each U_i. Clearly, f is an immersion with respect to this manifold structure.

Suppose that (X, f) satisfies (Im). Then Theorems 1 and 2 together show that there is a unique manifold structure on X such that f is an immersion. We call this structure on X the *inverse image structure on X relative to f* or simply the *induced structure* on X. We write X_f if we wish to make explicit the dependence of this structure on f.

We now give several applications of the above results.

A) Submanifolds

Suppose X is a subspace of Y (with the induced topology) and let

$$\imath : X \to Y$$

be the inclusion map. If $(X, \imath)$ satisfies (Im) we say that X is a *submanifold* of Y; note that this implies that X is locally closed in Y.

Let $x \in X$. One says that X is locally a submanifold of Y at x if the following equivalent conditions are satisfied:

1. $(X, \imath)$ satisfies (Im) at x.

2. There is an open neighborhood U of x in Y such that $U \cap X$ is a submanifold of U.

3. There exist a coordinate system $x_1, \ldots, x_n$ at x and an integer $p \leq n$ such that X is given by $x_1 = \cdots = x_p = 0$ in a neighborhood of x.

B) Local homeomorphisms

When f is a local homeomorphism, (X, f) satisfies (Im). In this case, the morphism $f : X_f \to Y$ is étale.

C) Inverse images of points

Let X and Y be manifolds, $f : X \to Y$ be a morphism, and $b \in Y$. Set $X_b = f^{-1}(b)$ and let $a \in X_b$. We shall study $X_b \subset X$ in a neighborhood of a.

Theorem. *The set X_b is locally a submanifold of X at a if any one of the following three conditions is satisfied:*

1. *f is a subimmersion in a neighborhood of a.*

2. *There exists a submanifold W of X such that:*

1) $W \subset X_b$.

2) $T_a W = \mathrm{Ker}(T_a X \xrightarrow{T_a f} T_b Y)$.

3. (Weil) *There exists a manifold Z, a point $c \in Z$, and a morphism $g : Z \to X$ such that:*

1) *For all $z \in Z$, $f \circ g(z) = b$.*
2) *$g(c) = a$.*
3) *The sequence $T_cZ \xrightarrow{T_c g} T_aX \xrightarrow{T_a f} T_bY$ is exact.*

Moreover, in each case, $T_a(X_b) = \operatorname{Ker}(T_aX \xrightarrow{T_a f} T_bY)$.

Proof. 1. The proof is an immediate consequence of the definition of a subimmersion.

2. We shall prove the following stronger statement: There exists an open neighborhood U of a in X such that $U \cap X_b = U \cap W$.

The statement is local so we may assume that X is an open neighborhood of $a = 0$ in k^m and that $X = W \times V$. Define $F : X \to W \times Y$ by the formula: $F(w, v) = (w, f(w, v))$. Then F is an immersion at 0 so by shrinking X we may also assume F injective. Then $X_b \subset F^{-1}(W \times \{b\}) = W \times \{0\} = W$.

3. We shall prove the following stronger statement: There exist open neighborhoods W of c in Z and U of a in X, a decomposition $W = W_1 \times W_2$, and a morphism $\phi : W_1 \to X$ such that:

a. ϕ is an isomorphism of W_1 into a submanifold ϕW_1 of X.

b. The map g factors as:

$$W_1 \times W_2 \xrightarrow{\mathrm{pr}_1} W_1 \xrightarrow{\phi} X .$$

c. $U \cap X_b = g(W)$.

In particular, this will show that g is a subimmersion at c.

The statement is local so we may assume that Z is an open neighborhood of $c = 0$ in k^p and that $Z = W_1 \times W_3$ where $T^{W_1}(g)$ is an isomorphism at 0 and $T^{W_3}(g)$ is zero at 0. Let $\phi = g|_{W_1}$. Then ϕ is an isomorphism at 0 so we may assume by shrinking W_1 that ϕ is an isomorphism of W_1 onto a submanifold of X. Then, by 1) and 3), ϕW_1 satisfies the hypotheses of part 2. Hence there is an open neighborhood U of a in X such that $U \cap X_b = U \cap \phi(W_1)$.

There is an open neighborhood W of 0 in $W_1 \times W_3$ such that $g(W) \subset U \cap X_b$. Then, $g : W \to \phi W_1$ and this map is a submersion at 0. Hence by shrinking W and W_1, we may find a product decomposition $W = W_1 \times W_2$ such that conditions a) and b) are both satisfied. Finally we can shrink U so that c) is also true. q.e.d.

D) Transversal submanifolds

Let X be a manifold, Y_1 and Y_2 be submanifolds of X, and $x \in Y_1 \cap Y_2$.

Theorem. *The following are equivalent:*

1. $T_xX = T_xY_1 + T_xY_2$.

2. *There is a chart* $c = (U, \phi, n)$ *at* x *such that:*

$$\begin{aligned}\phi U &= V_1 \times V_2 \times W \\ \phi(U \cap Y_1) &= V_1 \times \{0\} \times W \\ \phi(U \cap Y_2) &= \{0\} \times V_2 \times W \ .\end{aligned}$$

3. *There exists a coordinate system* $x_1, \ldots, x_n$ *at* x *and integers* $p, q \geq 0$ *with* $p + q \leq n$ *such that:*

Y_1 *is given by* $x_1 = \cdots = x_p = 0$ *in a neighborhood of* x,
Y_2 *is given by* $x_{p+1} = \cdots = x_{p+q} = 0$ *in a neighborhood of* x.

Proof. 2 $\iff$ 3 and 2 $\Rightarrow$ 1: Obvious.

1 $\Rightarrow$ 3: Since Y_1 and Y_2 are submanifolds of X, we can (after suitably shrinking X) find submersions

$$f_1 : X \to k^p \ , \qquad f_2 : X \to k^q$$

such that $Y_i = f_i^{-1}(0)$, $i = 1, 2$. Let $(x_1, \ldots, x_p)$ and $(x_{p+1}, \ldots, x_{p+q})$ be the components of f_1 and f_2. Assumption 1 implies that the map

$$(f_1, f_2) : X \to k^p \times k^q$$

is a submersion at x. This means that $(x_1, \ldots, x_{p+q})$ is a part of a coordinate system $(x_1, \ldots, x_n)$ at x. Hence 1 $\Rightarrow$ 3.

If Y_1 and Y_2 satisfy the equivalent conditions of the preceding theorem at x, we say that Y_1 and Y_2 are *transversal* at x.

Corollary. *Suppose* Y_1 *and* Y_2 *are transversal at* x. *Then:*

1. Y_1 *and* Y_2 *are transversal in a neighborhood of* x.
2. $Y_1 \cap Y_2$ *is locally a submanifold of* X *at* x.
3. $T_x(Y_1 \cap Y_2) = T_x Y_1 \cap T_x Y_2$.

E) Transversal morphisms

Consider a pair of morphisms $f_i : Y_i \to X$, $i = 1, 2$. Define

$$Y_1 \times_X Y_2 = \{ (y_1, y_2) \in Y_1 \times Y_2 : f_1(y_1) = f_2(y_2) \} \ .$$

This is called the *fibre product* of Y_1 and Y_2 over X. Let $p_i : Y_1 \times_X Y_2 \to Y_i$ be the restriction of pr_i to $Y_1 \times_X Y_2$, and let $f = f_1 \circ p_1 = f_2 \circ p_2$.

$$\begin{array}{ccc} Y_1 \times_X Y_2 & \xrightarrow{p_2} & Y_2 \\ {\scriptstyle p_1}\downarrow & \searrow^{f} & \downarrow{\scriptstyle f_2} \\ Y_1 & \xrightarrow[f_1]{} & X \end{array}$$

Let $(y_1, y_2) \in Y_1 \times_X Y_2$ and let $x = f(y_1, y_2)$. We say that f_1 and f_2 are *transversal at* $y = (y_1, y_2)$ if $T_x X = \operatorname{Im} T_{y_1} f_1 + \operatorname{Im} T_{y_2} f_2$.

Theorem. *Suppose f_1 and f_2 are transversal at y. Then:*

1. *f_1 and f_2 are transversal at points in a neighborhood of y in $Y_1 \times_X Y_1$.*
2. *$Y_1 \times_X Y_2$ is locally a submanifold of $Y_1 \times Y_2$ at y.*
3. *$T_y(Y_1 \times_X Y_2) = T_{y_1}(Y_1) \times_{T_x(X)} T_{y_2}(Y_2)$.*

Sketch of proof. Set $Y = Y_1 \times Y_2$ and $Z = Y_1 \times_X Y_2$. Let $\delta_i : Y \to Y \times X$ be $(1, f_i \circ \mathrm{pr}_i)$, $i = 1, 2$, and let $\delta = \delta_i|_Z$. Then deduce the theorem from:
a. δ_1 and δ_2 are isomorphisms of Y onto submanifolds of $Y \times X$.
b. $\delta_1(Y)$ and $\delta_2(Y)$ are transversal at $\delta(y)$.
c. $\delta(Z) = \delta_1(Y) \cap \delta_2(Y)$.
The details are left to the reader.

Remark. If one of the maps f_i is a submersion, then f_1 and f_2 are everywhere transversal.

F) Mixed transversality
If, in the situation of E), f_1 is an inclusion of a submanifold Y_1 into X, we also say that f_2 is *transversal over* Y_1 at y if f_1 and f_2 are transversal at y.

12. Construction of manifolds: quotients

Let X be a manifold and $R \subset X \times X$ be an equivalence relation. Let X/R be the set of equivalence classes of elements of X under R and let $p : X \to X/R$ be the projection. Give X/R the usual quotient topology, namely, let $\bar{U} \subset X/R$ be open if and only if $p^{-1}(\bar{U}) \subset X$ is open.

Theorem 1. *If there exists a manifold structure on X/R such that p is a submersion then this manifold structure is unique.*

Proof. By Lemma 2 of III.11, for all Z, $\mathrm{Mor}(X/R, Z)$ is determined by the manifold structure of X. Hence, by the Theorem of III.11 the manifold structure on X/R is unique.

When a manifold structure can be defined on X/R such that p is a submersion, then we give X/R this uniquely determined structure and say that X/R is a *quotient manifold* of X, or simply a manifold; the relation R is called a *regular* equivalence relation on X.

Theorem 2 (Godement). *The following are equivalent:*

1. *X/R is a manifold, that is, R is regular.*
2. 1) *R is a submanifold of $X \times X$.*

2) $\mathrm{pr}_2 : R \to X$ *is a submersion.*

Proof. $1 \Rightarrow 2$:

$$\begin{array}{ccc} R & \xrightarrow{\mathrm{pr}_2} & X \\ {\scriptstyle \mathrm{pr}_1}\downarrow & & \downarrow{\scriptstyle p} \\ X & \xrightarrow[p]{} & X/R \end{array}$$

The set R is equal to $X \times_{X/R} X$. Since p is a submersion, R is a submanifold of $X \times X$, cf. III.11, n° 2, E). Moreover, if $(x, y) \in R$ and $z = p(x) = p(y)$, one has:

$$T_z(R) = T_x(X) \times_{T_z(X/R)} T_y(X) .$$

This formula implies that $T_z(R) \to T_y(X)$ is surjective, hence the restriction of pr_2 to R is a submersion.

$2 \Rightarrow 1$: We shall give a sequence of six lemmas which together yield $2 \Rightarrow 1$.

Suppose U is a subset of X. Set $R_U = R \cap (U \times U)$. Also, recall that U is said to be *saturated* with respect to R if $U = p^{-1}p(U)$.

Lemma 1. *Assume* $X = \bigcup_{i \in I} U_i$ *where, for* $i \in I$, U_i *is an open saturated subset of* X *such that* U_i/R_{U_i} *is a manifold. Then* X/R *is a manifold.*

Proof. By hypothesis, for $i \in I$, $U_i \to U_i/R_{U_i}$ is a submersion. Hence, for $i, j \in I$, the manifold structures induced on $(U_i \cap U_j)/R_{(U_i \cap U_j)}$ by U_i/R_{U_i} and U_j/R_{U_j} agree (Theorem 1). Hence there is a unique manifold structure on X/R compatible with the given structure on U_i/R_{U_i}. Finally, p is a submersion since $p|_{U_i}$ is a submersion for all i.

Lemma 2. *The map* p *is open, that is,* U *open in* $X \Rightarrow p^{-1}p(U)$ *open in* X.

Proof. We have that $p^{-1}p(U) = \mathrm{pr}_2(U \times X \cap R)$ which is open if U is open because pr_2 is a submersion (III.11, Lemma 2).

Lemma 3. *Let* U *be open in* X *and suppose that* $p^{-1}p(U) = X$ *and that* U/R_U *is a manifold. Then* X/R *is a manifold.*

Proof. The canonical map $\alpha : U/R_U \to X/R$ is bijective. Hence, if we show that $\beta = \alpha^{-1}p : X \to U/R_U$ is a submersion, we will obtain by transporting the structure of U/R_U to X/R that X/R has a manifold structure such that p is a submersion. Consider the following commutative diagram:

$$\begin{array}{ccccc} & & U \times X \cap R & & \\ & {\scriptstyle \mathrm{pr}_1}\swarrow & & \searrow{\scriptstyle \mathrm{pr}_2} & \\ U & & & & X \\ & {\scriptstyle \beta|_U}\searrow & & \swarrow{\scriptstyle \beta} & \\ & & U|_{R_U} & & \end{array} .$$

Then $(\beta|_U)\circ(\mathrm{pr}_1) = \beta\circ(\mathrm{pr}_2)$ is a submersion. Hence, since pr_2 is a submersion, β is a morphism and in fact a submersion (III.11, Lemma 2).

Combining Lemmas 1, 2, and 3, we obtain immediately:

Lemma 4. *Assume $X = \bigcup_{i\in I} U_i$ where, for $i \in I$, U_i is an open subset of X such that U_i/R_{U_i} is a manifold. Then X/R is a manifold.*

The effect of Lemma 4 is to make the construction of a manifold structure on X/R such that p is a submersion into a local problem. In Lemmas 5 and 6, we show that the local problem is solvable, that is, for $x_0 \in X$, there is a neighborhood U of x_0 in X such that U/R_U has a manifold structure such that $U \to U/R_U$ is a submersion.

Lemma 5. *Let $x_0 \in X$. Then there exists an open neighborhood U of x_0, a manifold W of U, and a morphism $r : U \to W$ such that if $u \in U$ then $r(u)$ is the unique point of W equivalent to u* mod *R.*

Proof. Let N be the set of tangent vectors $\xi \in T_{x_0}(X)$ such that $(\xi, 0) \in T_{x_0,x_0}(R)$. Choose a submanifold W' of X such that $x_0 \in W'$ and $K = T_{x_0}W'$ is a complementary subspace to N in $T_{x_0}X$. Then define $\Sigma = (W' \times X) \cap R$.

We contend that:

1. Σ is a submanifold of R.

2. $\mathrm{pr}_2 : \Sigma \to X$ is étale at (x_0, x_0).

The first assertion follows since $\Sigma = \psi^{-1}(W')$ where ψ denotes the submersion $\mathrm{pr}_1 : R \to X$. Note that we have used the results of III.11, E) and applied the hypothesis that pr_2 is a submersion together with the fact that R is an equivalence relation which shows that pr_1 is also a submersion.

Next, $\mathrm{Ker}(T(\mathrm{pr}_2))$ at (x_0, x_0) is $N \cap K = 0$. Hence $T(\mathrm{pr}_2)$ is injective. On the other hand, let $\eta \in T_{x_0}X$ and choose $\xi \in T_{x_0}X$ such that $(\xi, \eta) \in T_{x_0,x_0}R$. Write $\xi = \xi_1 + \xi_2$ where $\xi_1 \in N$ and $\xi_2 \in K$. Then, it is also true that $(\xi_2, \eta) \in T_{x_0,x_0}R$ since $N \subset T_{x_0,x_0}R$. But (ξ_2, η) then belongs to $T_{x_0}W' \times T_{x_0}X \cap T_{x_0,x_0}R = T_{x_0,x_0}\Sigma$ and this element also maps onto η. Hence $T(\mathrm{pr}_2)$ is surjective.

It follows that there exists a pair of open neighborhoods U_1 and U_2 of x_0 such that $\mathrm{pr}_2 : \Sigma \cap (U_1 \times U_1) \to U_2$ is an isomorphism. Let f denote the inverse. Then f must have the form: $f(x) = (r(x), x)$. Notice that $U_2 \subset U_1$ and that if $x \in U_2 \cap W'$ then $r(x) = x$. The last statement follows from the fact that (x, x) and $(r(x), x)$ are two points in $\Sigma \cap (U_1 \times U_1)$ with the same image in U_2 and hence are equal.

Finally, set $U = \{ x : x \in U_2 \text{ and } r(x) \in U_2 \cap W' \}$ and set $W = U \cap W'$. We contend that U, W and r are as required in the statement of the lemma. We must show that:

1. $r(U) \subset W$.

2. $r(x)$ is the only element of W equivalent to x, for $x \in U$.

To prove 1, we must show that, for $x \in U$, $r(x) \in U$, that is, that $r(x) \in U_2$ which is obvious and that $r(r(x)) \in U_2 \cap W'$. The last statement follows since $r(r(x)) = r(x) \in U_2 \cap W'$. To prove 2, we simply note that there is exactly one point in $R \cap (W \times U)$ mapping by pr_2 onto x, namely, $(r(x), x)$.

This completes the proof of the lemma.

Lemma 6. *If (U, W, r) satisfy the conditions described in Lemma 5, then U/R_U is a manifold.*

Proof. The morphism $r : U \to W$ has a right inverse (the inclusion of W into U); hence it is a submersion. In the commutative diagram:

$$\begin{array}{ccc} U & \xrightarrow{r} & W \\ & \searrow \quad \swarrow_{\alpha} & \\ & U/R_U & \end{array}$$

the map α is a bijection. Transporting the manifold structure of W to U/R_U, we have the lemma. q.e.d.

Remark. If R is regular, X/R is Hausdorff if and only if R is closed in $X \times X$ (this follows from Lemma 2 above).

Exercises

1. Let G be a finite group of automorphisms of a manifold X, and let X^G be the set of fixed points of G. Assume the order of G is prime to the characteristic of k. Show:

   a) If $x \in X^G$, there is a system of local coordinates at x with respect to which G acts linearly.

   b) X^G is a submanifold of X, and, if $x \in X^G$, $T_x(X^G)$ is equal to $T_x(X)^G$.

2. Assume k is a perfect field of characteristic $p \neq 0$. Let X be a manifold over k. Show that there exists on the topological space X a unique structure of manifold (denoted by X^p) with the following property:

   If Y is any manifold, $\mathrm{Mor}(X^p, Y)$ is equal to the set of morphisms $f : X \to Y$ such that $T_x(f) = 0$ for all $x \in X$.

   A map $f : X \to k$ is an X^p-morphism if and only if its p-th root is an X-morphism.

   Show the existence of $X^{p^{-1}}$ such that $(X^{p^{-1}})^p = X$, and define inductively X^q for $q = p^n$, with $n \in \mathbf{Z}$. Show that $\mathrm{Mor}(X^q, Y^q) = \mathrm{Mor}(X, Y)$. One has $X^q = X$ if and only of $q = 1$ or X is discrete (i.e., of dimension 0).

3. Assume k is locally compact ultrametric; let A_v be its valuation ring, $\mathfrak{m}_v = \pi A_v$ the maximal ideal of A_v, $k(v) = A_v/\mathfrak{m}_v$, and $q = \mathrm{Card}\, k(v)$. Let $B = (A_v)^N$ be the unit ball of some dimension N, and put $B_n = (A_v/\pi^n A_v)^N$, so that $B = \varprojlim B_n$. Let X be a non-empty submanifold of B; assume X is everywhere of dimension d. Let X_n be the image of X in B_n, and $c_n = \mathrm{Card}(X_n)$. Show:

   a) There exist $n_0 \geq 0$ and $A > 0$ such that:
$$c_n = A \cdot q^{nd} \qquad \text{for } n \geq n_0.$$

   b) Let $a \in \mathbf{Z}/(q-1)\mathbf{Z}$ be the invariant of X defined in Appendix 2 (assuming now that $d \geq 1$); one has $A \equiv a \bmod (q-1)$.

4. Let X be a manifold, X_i be submanifolds of X, and $x \in \bigcap X_i$. Assume that the $T_x(X_i)$ are linearly independent in $T_x(X)$ (i.e., the sum of the $T_x(X_i)$ is a direct sum). Show that there exists a chart $c = (U, \phi, n)$ on X, with $x \in U$, such that $\phi(U|_{X_i})$ is the intersection of $\phi(U)$ with a linear subvariety of k^n.

5. Let f_i $(i = 1, 2) : X_i \to X$ be transversal morphisms, and let
$$p_i : X_1 \times_X X_2 \to X_i$$
be the projection morphisms. Show that, if f_1 is a submersion (resp. an immersion, a subimmersion), the same is true for p_2.

6. Let $f : X \to Y$ be a morphism. Assume f is open and the characteristic of k is zero. Show that the set of points of X where f is a submersion is dense in X.

Appendix 1. A non-regular Hausdorff manifold

An example of a Hausdorff manifold over an ultrametric field k which has a point which does not have a fundamental system of open and closed neighborhoods. The example is due to George Bergman.

Let k be a complete ultrametric field and let A be its valuation ring. Suppose there exists $x \in A$ such that $x \neq 0$ and A/xA is infinite. Then we contend that A is analytically isomorphic to $A - \{0\}$. To show this, we shall show that A and $A - \{0\}$ may both be represented as the disjoint union of the same cardinal number of copies of A. First note that if μ is a positive integer then the cosets of $x^\mu A$ are isomorphic to A. Then note that A is the disjoint union of the cosets of xA while $A - \{0\}$ is the disjoint union of the following collection of cosets of $x^\mu A$ where μ ranges over the positive integers:

1. The cosets of xA excepting xA itself.
2. The cosets of $x^2 A$ in xA excepting $x^2 A$ itself.

$\vdots$

μ. The cosets of $x^\mu A$ in $x^{\mu-1} A$ excepting $x^\mu A$ itself.

Since A/xA is infinite, it is clear that both sets of cosets which we have described have the same cardinality.

We may view the above construction as smoothly attaching a point P to the ball A: $A \subset A \cup \{P\} \simeq A$, and P is the point 0 in the latter copy of A. This attaching process has three important properties:

1. $A \cup \{P\}$ is a Hausdorff analytic manifold.

2. P is in the closure of A.

3. P is not in the closure of any coset of the maximal ideal $\mathfrak{m}$ of A.

This last property is a consequence of the fact that 0 is "far away" from any of the cosets we have used to describe $A - \{0\}$ as a disjoint union of copies of A.

We are now going to do this attaching process a countable number of times. Attach in the above manner a point P_0 to A. Since $xA \approx A$, attach a point P_1 in the above manner to xA. Property 3 says that the point P_1 is "far away" from P_0 so that we again have a Hausdorff manifold. Suppose now that we have attached points $P_0, \ldots, P_\mu$ to $A, \ldots, x^\mu A$. Then attach a point $P_{\mu+1}$ to $x^{\mu+1}A$. Now pass to the limit. We give the limit the topology such that each of the subsets X, $P_0, \ldots, P_\mu$ is open and has its own original topology.

Since the points we have attached are "far away" from each other, it is clear that the manifold X we obtain in the limit is Hausdorff. However the point $0 \in A$ does not have a fundamental system of neighborhoods which are open and closed. Indeed the powers $\{x^\mu A\}$ are a fundamental system of neighborhoods of 0. If we had a fundamental system of neighborhoods of 0 which was open and closed we could find one such neighborhood U contained in A. Then find $x^\mu A \subset U$. The closure of $x^\mu A$ contains $P_\mu \notin A$. Contradiction.

Remark. The reader should verify that there exists $x \in A$ such that $x \neq 0$ and A/xA is infinite if and only if one of the following two conditions is satisfied:

1. The residue field of k is infinite.

2. The valuation of k takes on a non-discrete set of values.

The only ultrametric fields not satisfying one of these conditions are the finite extensions of the p-adic fields $\mathbf{Q}_p$ and the fields $F((X))$ where F is a finite field.

Appendix 2. Structure of p-adic manifolds

We shall use the notion of disjoint union to study manifolds in the case when k is *locally compact* and *ultrametric*. We let $n \in \mathbf{Z}$, $n \geq 0$, and we assume that X is everywhere of dimension n. We also assume that X is Hausdorff, and non-empty.

Lemma 1. *Let $r \in \mathbf{R}$, $r > 0$, and let $x \in k^n$. Then $B(r)(x)$ is compact and open. Hence every ball in X is compact and open.*

Proof.
1. Compactness:

Since k is locally compact, there is a positive real number ε such that any ball of radius $s < \varepsilon r$ about x in k^n is contained in a compact neighborhood of x. Since these balls are closed in k^n, they are compact. Since the absolute value on k is non-trivial, we may choose $\alpha \neq 0, \in k$, such that $|\alpha| < \varepsilon$. Then the transformation $f(y) = x + \alpha(y - x)$ is a topological isomorphism of $B(r)(x)$ onto $B(|\alpha|r)(x)$. Hence $B(r)(x)$ is compact.

2. Openness:

We contend that if $y \in B(r)(x)$ then $B(r)(y) = B(r)(x)$ so that $B(r)(x)$ is a neighborhood of y. Since $x \in B(r)(y)$, it suffices by symmetry to show $B(r)(y) \subset B(r)(x)$. Let $z \in B(r)(y)$. Then:

$$|z - x| \leq \max(|z - y|, |y - x|) \leq r .$$

Thus, $z \in B(r)(x)$ as desired. Note that we have used here the fact that k is ultrametric.

Remark. An analogous argument shows that, if B_i are balls of radius r_i, $i = 1, 2$, and $r_1 \leq r_2$, then B_1 is contained in B_2 or is disjoint from B_2.

Lemma 2. *Let U be a closed and open set of a ball B in k^n. Then there is a positive radius r smaller than the radius of B such that U is the disjoint union of a finite number of balls of radius r.*

Proof. Let $V = B - U$. Then $\{U, V\}$ is an open covering of the compact metric space B. Hence there is a radius r less than the radius of B such that, for all $x \in B$, the ball of radius r about x in B is contained in either U or V. By the preceding remark, we see that a ball of radius r in B is a ball of radius r in k^n. Hence U is the union of balls of radius r in k^n. The union is disjoint by the preceding remark and therefore finite since U is compact.

Remark. From the lemma, we see that if B is a ball in X and U is an open and closed set in B, then U is the disjoint union of a finite number of balls in X.

Theorem 1. *The following are equivalent:*

1. *X is paracompact* (Bourbaki, *TG.* I. 69).
2. *X is the disjoint union of balls.*

Proof. 2 $\Rightarrow$ 1: A disjoint union of compact spaces is paracompact.

1 $\Rightarrow$ 2: We shall first show that X has a locally finite covering by balls. We know that X has a covering $\{U_\lambda\}_{\lambda \in L}$ by balls. Choose a locally finite open refinement $\{V_\mu\}_{\mu \in M}$ of this covering. Then choose a locally finite closed refinement $\{W_\nu\}_{\nu \in N}$ of this covering. Let $\phi : M \to L$ and $\psi : N \to M$ be such that $V_\mu \subset U_{\phi(\nu)}$ and $W_\nu \subset V_{\psi(\nu)}$. Let $\nu \in N$. Then:

$$W_\nu \subset V_{\psi(\nu)} \subset U_{\phi\psi(\nu)} \, .$$

Since W_ν is closed and $U_{\phi\psi(\nu)}$ is compact, W_ν is compact. Then, since $V_{\phi(\nu)}$ is open, we may cover W_ν by a finite number of balls $B_{\nu,i}$, $i \in I_\nu$, such that $B_{\nu,i} \subset V_{\phi(\nu)}$ for all i. Then the covering $\{B_{\nu,i}\}_{\nu\in N, i\in I_\nu}$ consists of balls and is locally finite since any ball $B_{\nu,i}$ meets at most finitely many V_μ and hence only finitely many $B_{\nu',i'}$.

We will now simplify notation and let $\{U_i\}_{i\in I}$ denote the locally finite covering of X by balls which we have obtained above. Then each U_i is open and compact and meets only finitely many U_j. Let $F(I)$ denote the finite subsets of I. Then, if $J \in F(I)$, define:

$$U_J = \bigcap_{i\in J} U_i \cap (X - \bigcup_{j\notin J} U_j) \, .$$

The set $(X - \bigcup_{j\in J} U_j)$ is open and compact. Indeed, if $J = \emptyset$, this is obvious, while if $i \in J$, then this set is the finite intersection of open and compact sets, namely, $\bigcap_{j\notin J}(U_i - U_j)$, where the j's may be restricted to the finite set of indices for which $U_j \cap U_i \neq \emptyset$. It follows that when U_J is non-empty then U_J is an open, compact subset of a ball, hence a finite unions of balls. However, by definition, the U_J, $J \in F(I)$ are disjoint. Thus we have the theorem using the covering $\{U_J\}_{J\in F(I)}$.

Theorem 2. *Let q be the number of elements of the residue field of k. Suppose X is compact, non-empty, and everywhere of the same dimension $d \geq 1$. Then:*

1. *X is the disjoint union of a finite number of balls.*

2. *The number of balls in a decomposition of X into a disjoint union of a finite number of balls is well determined* mod $(q-1)$.

(Hence, such an X is determined, up to an isomorphism, by an element of $\mathbf{Z}/(q-1)\mathbf{Z}$.)

Sketch of proof. 1. Follows immediately from Theorem 1 and the compactness of X.

2. We shall state a sequence of reduction steps and then shall prove the statement that the theorem is finally reduced to. Each of the reduction steps is based on the fact that one may divide a ball into q^i balls, where i is a positive integer, without disturbing congruences mod $(q-1)$.

Suppose X is given with two decompositions $\{U_i\}_{i\in I}$ and $\{V_j\}_{j\in J}$ where I and J are finite and $\{U_i\}$ and $\{V_j\}$ are made up of disjoint balls. Then we want to show that $\mathrm{Card}(I) \equiv \mathrm{Card}(J) \bmod (q-1)$.

Step 1: Reduce to the case when $\{U_i\}_{i\in I}$ is a refinement of $\{V_j\}_{j\in J}$.

Step 2: Reduce to the case when $X = V_j$ and $J = \{j\}$. Then we have the following explicit situation:

a. X is a ball in k^n.
b. U_i is a ball in k^n for $i \in I$.
c. There exist analytic isomorphisms ϕ_i, $i \in I$, of U_i into X such that X is the disjoint union of $\{\phi_i U_i\}$.

Step 3: Reduce to the case when each ϕ_i is given by a convergent power series.

Step 4: Reduce to the case $\phi_i = L_i \circ \psi_i$ where L_i is a linear isomorphism and ψ_i is an isomorphism of a ball onto a ball. We can then assume that $\psi_i = L_i$.

Step 5: We contend that it suffices to prove that there are radii r_i such that, for any radii $s_i \leq r_i$, $L_i U_i$ is the disjoint union of radius s_i and such that the number of such balls is a power of q. For, if this is so, we take $r = \min(r_i)$. Then X is decomposed in q^m balls of radius r while each $L_i U_i$ is decomposed into q^{m_i} balls of radius r. Then:

$$1 \equiv q^m = \sum_{i\in I} q^{m_i} \equiv \sum_{i\in I} 1 \mod (q-1)\,.$$

This is precisely what has to be proved in this special case.

We are therefore reduced to showing that if U is a ball and L is a linear isomorphism then there is a radius r such that:

1. If $0 < s \leq r$, then LU is the disjoint union of balls of radius s.

2. The number of such balls is a power of q.

By translation and multiplication by scalars, we may assume $U = A_v$ and $L \in M_n(A_v)$, where A_v is the valuation ring of k. We let $\mathfrak{m}_v$ denote the maximal ideal of A_v and we note that number of cosets of $\mathfrak{m}_v^\nu$, where μ is a positive integer, is equal to the number of elements in $A_v/\mathfrak{m}_v^\mu$ which is exactly q^μ, where q is the number of elements in $A_v/\mathfrak{m}_v$.

The existence of a radius r satisfying 1 is guaranteed by Lemma 2. We contend that this radius also satisfies 2. Indeed let $0 < s \leq r$ and let h be the number of balls of radius s in U. Now, A_v^n is the disjoint union of a finite number h' of translates of U, so that hh' is the number of balls of radius s in A_v^n. Let μ be the positive integer such that the ideal $\mathfrak{m}_v^\mu$ is precisely the ball of radius s in A_v. Then $(\mathfrak{m}_v^\mu)^n$ is the ball of radius s in A_v^n about 0. Hence there are $(q^\mu)^n$ balls of radius s in A_v^n. However, h' is also a power of q. This follows since $h' = \mathrm{Card}(A_v^n/U)$ and since A_v^n/U is a torsion module over A_v and hence a direct sum of modules of the form $A_v/\mathfrak{m}_v^{\mu_i}$ each of which has cardinality equal to a power of q. Thus, finally, h is a power of q. q.e.d.

Remark. For a different proof of Theorem 2 (using integration of differential forms) see "Topology", vol. 3 (1965), 409–412.

Appendix 3. The transfinite p-adic line

A propos of Theorem 1, there exist non-paracompact Hausdorff manifolds over any locally compact ultrametric field k. We give here an example of such a manifold which is due to George Bergman.

We shall construct a directed system $\{X_\gamma\}$ of spaces indexed by the elements of the first uncountable ordinal and our example of a non-paracompact manifold will be given by $X = \varinjlim X_\gamma$.

The manifolds X_γ will all be taken to be copies of the valuation ring A of k. We shall define the maps $X_\delta \to X_\gamma$ for $\delta < \gamma$ by induction on γ.

1) $\gamma = 0$.

The condition $\delta < \gamma$ is vacuous in this case.

2) $\gamma = \gamma' + 1$, for some ordinal γ'.

Let π be a fixed generator of the maximal ideal $\mathfrak{m}$ of X. Let $X_{\gamma'} \to X_\gamma$ be multiplication by π.

For arbitrary $\delta < \gamma$, let $X_\delta \to X$ be the composite $X_\delta \to X_{\gamma'} \to X_\gamma$.

3) γ is an initial ordinal.

Let $Y_\gamma = \lim_{\delta<\gamma} X_\delta$. Then Y_γ is the union of the countable family of open, compact subspaces X_δ $(\delta < \gamma)$ and is therefore paracompact. By Theorem 1, it is the disjoint union of balls. The number of such balls must be countable in number since a disjoint union is locally finite and only finitely many elements of any locally finite covering can meet any given X_δ. Since $A - \{0\}$ is also the union of a countable number of balls, we may choose an analytic isomorphism $\phi_\gamma : Y_\gamma \to A - \{0\}$. Then, for $\delta < \gamma$, the map $X_\delta \to X_\gamma$ is defined to be the composite: $X_\delta \longrightarrow Y_\gamma \xrightarrow{\phi_\gamma} A - \{0\} \subset A = X_\gamma$. The inductive definition of the maps $X_\delta \to X_\gamma$ for $\delta < \gamma$ is now complete.

The space X so constructed has the following two properties:

(1) Any denumerable family (K_n) of compact subsets of X is contained in a compact set.

(2) X is not compact.

Proof of (1). Since $K_n = \bigcup(K_n \cap X_\gamma)$, and X_γ is open, there exists a γ_n with $K_n \subset X_{\gamma_n}$; choosing γ such that $\gamma_n \leq \gamma$ for all n, we have $K_n \subset X_\gamma$, and X_γ is compact.

Proof of (2). Follows from $X_\gamma \neq X$ for all γ.

We leave to the reader the verification of the fact that a locally compact space X with properties (1) and (2) is not paracompact.

Chapter IV. Analytic Groups

We denote by k a field complete with respect to a non-trivial absolute value.

1. Definition of analytic groups

Let G be a topological group and an analytic manifold over k. Then G is said to be an *analytic group* or a *Lie group* over k if the following conditions are satisfied:

1. The map $(x,y) \mapsto xy$ of $G \times G$ into G is a morphism.
2. The map $x \mapsto x^{-1}$ of G into G is a morphism.

Remarks. 1) Suppose G is an analytic group. Then:

a. G is Hausdorff.
b. G is metrizable.
c. G is complete for the left or right uniform structures.

Indeed, a) follows since a topological group is Hausdorff if and only if the intersection of the neighborhoods of the identity equals $\{e\}$. See Bourbaki, *TG.*, III.5. The second condition is satisfied in this case since G is locally isomorphic to an open subset of k^n for some integer n.

Statement b) is a consequence of the fact that G is Hausdorff and that e has a denumerable fundamental system of neighborhoods. See Bourbaki, *TG.*, IX.23.

To show statement c), it suffices to consider only the right uniform structure. Furthermore, it suffices to show that there is a neighborhood V of e which is complete in the induced uniform structure. See Bourbaki, *TG.*, III.22. We construct such a neighborhood V of e as follows. Let (U, ϕ, n) be a chart at e such that $\phi(e) = 0$ and let V_1 be a neighborhood of e such that $V_1 \cdot V_1 \subset U$. Then the law of composition on V_1 is induced via ϕ from an analytic map $F : \phi V_1 \times \phi V_1 \to \phi U$. For $\bar{y} \in \phi V_1$, $F(\bar{y}, 0) - F(0,0) = \bar{y} - 0 = \bar{y}$. Then, since F is analytic, there is a closed neighborhood V of e in V_1 such that for $(\bar{y}, \bar{x}) \in \phi V \times \phi V$ we have

$$\tfrac{1}{2}|\bar{y}| \leq |F(\bar{y}, \bar{x}) - F(0, \bar{x})| \leq 2|\bar{y}| \ .$$

We shall show that V is complete by showing that the uniform structure on V agrees with the uniform structure induced via ϕ from the uniform structure on ϕV given by the additive structure of k^n. Now, a fundamental system of entourages of the uniform structure of V given by sets of the form $V_W \subset V \times V$ where:

1. W is a neighborhood of e in V and ϕW is a ball of radius ε about 0.
2. $V_W = \{\, (w \cdot x, x) : x \in V,\ w \in W,\ w \cdot x \in V \,\}$.

On the other hand, let $N_\delta = \{\, (\bar{y}, \bar{x}) \in \phi V \times \phi V : |\bar{y} - \bar{x}| \leq \delta \,\}$ where $\delta > 0$. The sets N_δ form a fundamental system of entourages for the uniform structure on ϕV induced from the additive structure of k^n. Now, with W as above:

$$N_{\varepsilon/2} \subset (\phi \times \phi)V_W \subset N_{2\varepsilon} \, .$$

Hence the two uniform structures on V agree. Since the uniform structure induced by the additive structure of k^n is complete because V is closed, statement c) is proved.

Notice. We have shown that the left or right uniform structures locally agree with the uniform structures induced by charts.

2) Concerning the axioms of analytic groups:
a. Axiom 1 implies that for fixed $x \in G$ the map $y \mapsto xy$ is an isomorphism (for the manifold structure of G).
b. Axiom 1 implies Axiom 2.
c. Axiom 2 implies that the map $x \mapsto x^{-1}$ is an isomorphism.

Indeed, let $\phi : G \times G \to G$ denote the map $(x, y) \mapsto xy$, let $\phi_x : G \to G$ denote the map defined by $\phi_x(y) = \phi(x, y)$, and let $\psi : G \to G$ denote the map $x \mapsto x^{-1}$. Let $T^1\phi$ and $T^2\phi$ be the first and second partial derivatives of ϕ (see Chap. 3, §8).

Then, statement a) is a consequence of the fact that ϕ_x is the composite of the morphism $y \mapsto (x, y)$ of G into $G \times G$ with ϕ which shows that ϕ_x is a morphism and the fact that ϕ_x has an inverse, namely, $\phi_{x^{-1}}$. Note that $T_y\phi_x : T_yG \to T_{xy}G$ may be identified with $T^2_{x,y}\phi : T_yG \to T_{xy}G$. In particular, $T^2\phi$ is an isomorphism.

Statement b) is shown as follows. Consider the morphism

$$\theta : G \times G \to G \times G$$

defined by $\theta(x, y) = (x, xy) = (x, \phi(x, y))$. Then θ is bijective and étale at each point (x, y) of $G \times G$. Indeed, at (x, y), $T\theta$ has the form:

$$T\theta = \begin{pmatrix} T\,\mathrm{Id}_G & 0 \\ T^1\phi & T^2\phi \end{pmatrix} .$$

Thus, $T\theta$ is an isomorphism. It follows that θ is an isomorphism. Let $\sigma = \theta^{-1}$. Then, for all $x \in G$, $\sigma(x, e) = (x, x^{-1}) = (x, \psi x)$. Hence ψ is a morphism.

Statement c) is a consequence of the fact that $\psi^2 = 1$ which shows that ψ has an inverse and is therefore an isomorphism.

2. Elementary examples of analytic groups

1) General linear groups

Let R be an associate algebra with unit which is finite dimensional over k. The *general linear group* over R is the group $G_m(R)$ of invertible elements of R. We contend that $G_m(R)$ is an analytic group which is open as a subset of R. To show that $G_m(R)$ is open in R it suffices to show that there is a neighborhood of 1 contained in $G_m(R)$. Now, there exists an open neighborhood U of 0 in R such that for $x \in U$ the series $\sum x^n$ converges. It follows that

$$V = \{1 - x : x \in U\} \subset G_m(R)$$

and V is a neighborhood of 1. To show that $G_m(R)$ is an analytic group it remains to show that multiplication is a morphism. This follows since multiplication in R is bilinear.

In the special case where R is the endomorphism ring $E(V)$ of a finite dimensional vector space V over k, we call $G_m(R)$ the general linear group of V and denote it by $\mathrm{GL}(V)$. When $V = k^n$ we write $\mathrm{GL}(V) = \mathrm{GL}(n,k) = \mathrm{GL}_n(k)$. An element $\alpha \in \mathrm{GL}(n,k)$ may be represented as an n by n invertible matrix $\alpha = (\alpha_{ij})$. Hence $\mathrm{GL}(n,k)$ is called the *general linear group* of n by n matrices over k.

Suppose now that k is ultrametric and that A is the valuation ring of k. Then, for $\alpha = (\alpha_{ij}) \in \mathrm{GL}(n,k)$, the following are equivalent:

1. α defines an automorphism of A^n.
2. a. The coefficients α_{ij} of α lie in A.
   b. The determinant of α is a unit in A.

Let $\mathrm{GL}(n,A)$ denote the set of $\alpha \in \mathrm{GL}(n,k)$ satisfying the above conditions. Then, by condition 2, $\mathrm{GL}(n,A)$ is an open and closed subset of the set of n by n matrices with coefficients in A. Hence, in particular, $\mathrm{GL}(n,A)$ is open and closed in $E(k^n)$. By condition 1, $\mathrm{GL}(n,A)$ is a group. Hence, we have that $\mathrm{GL}(n,A)$ is an analytic group. We call $\mathrm{GL}(n,A)$ the general linear group of n by n matrices over A.

Suppose further that k is locally compact. Then $\mathrm{GL}(n,A)$ is a compact open subgroup of $\mathrm{GL}(n,k)$. In Appendix 1, we shall show:

Theorem. $\mathrm{GL}(n,A)$ *is a maximal compact subgroup of* $\mathrm{GL}(n,k)$ *and, if* G *is a maximal compact subgroup of* $\mathrm{GL}(n,k)$*, then* G *is a conjugate of* $\mathrm{GL}(n,A)$.

2) Induced analytic groups

Let G be an analytic group, H a topological group, and $\imath : H \to G$ a continuous homomorphism. Suppose that $(H,\imath)$ satisfies condition (Im) of Chap. 3, §11. Then H is a manifold with its induced structure. We contend that H is an analytic group. Indeed, let ϕ_G and ϕ_H denote the multiplication maps in G and H respectively. Then the following diagram is commutative:

$$\begin{array}{ccc} H \times H & \xrightarrow{\phi_H} & H \\ {\scriptstyle \imath\times\imath}\downarrow & & \downarrow{\scriptstyle \imath} \\ G \times G & \xrightarrow{\phi_G} & G \end{array}$$

Then $\phi_G \circ (\imath \times \imath)$ is a morphism; hence ϕ_H is a morphism since $\imath$ is an immersion. Therefore H is an analytic group.

Remarks. 1) To verify that $(H,\imath)$ satisfies (Im), it suffices to verify that $(H,\imath)$ satisfies (Im) at e_H. Indeed, suppose that $(H,\imath)$ satisfies (Im) at e_H and that

$h \in H$ and $g = \imath(h)$. Let $\phi : H \to H$ and $\psi : G \to G$ be defined by $\phi(x) = h^{-1}x$ and $\psi(y) = gy$. Then, $\phi(h) = e_H$, $\psi(e_G) = g$, and $\imath = \psi \circ \imath \circ \phi$. Since ψ is an analytic isomorphism and $\imath$ satisfies (Im) at e_H, $\psi \circ \imath$ satisfies (Im) at e_H. Then, since ϕ is a homeomorphism, $\imath$ satisfies (Im) at h.

2) We know, in particular, that $(H, \imath)$ satisfies (Im) when $\imath$ is a local homeomorphism (Chap. 3, §11, n°2, B). If, moreover, $\imath$ is surjective and $k = \mathbf{R}$ or $\mathbf{C}$, we say that H is a *covering group* of G.

3) Group submanifolds

Suppose G is an analytic group and H is a subgroup of G which is at the same time a submanifold of G. Then H is an analytic group. This is a special case of 2) since the inclusion $\imath : H \to G$ is a continuous homomorphism which is an immersion. We say in this case that H is a *group submanifold* of G.

Remark. Suppose that H is a group submanifold of G. Then H is closed in G. Indeed, this follows from:

1. A submanifold is locally closed in the manifold in which it lies.

2. A locally closed subgroup of a topological group is closed. See Bourbaki, *TG.*, III.7.

3. Group chunks

A *topological group chunk* is a topological space X together with a distinguished element $e \in X$, an open neighborhood U of e in X, and a pair of maps $\phi : U \times U \to X$ and $\psi : U \to U$ such that:

1. For some neighborhood V_1 of e in U, $x \in V_1$ implies that
$$x = \phi(x, e) = \phi(e, x) .$$

2. For some neighborhood V_2 of e in U, $x \in V_2$ implies that
$$e = \phi(x, \psi x) = \phi(\psi x, x) .$$

3. For some neighborhood V_3 of e in U, $\phi(V_3 \times V_3) \subset U$ and, for all x, y, z in V_3, $\phi(x, \phi(y, z)) = \phi(\phi(x, y), z)$.

We say that we have a *strict* group chunk if the equations in 1, 2, and 3 hold whenever both sides are defined. We can always obtain a strict group chunk from a group chunk by shrinking the open neighborhood U.

We shall often write $\phi(x, y) = xy$ and $\psi(x) = x^{-1}$ if no confusion is possible.

Let X and Y be group chunks. A *local homomorphism* $f : X \dashrightarrow Y$ is a continuous map $f : U \to Y$ where U is a neighborhood of e_X, such that $f(e_X) = e_Y$ and $f(xy) = f(x)f(y)$ in a neighborhood of e_X.

Two local homomorphisms $f, f' : X \dashrightarrow Y$ are called *equivalent* if they agree in some neighborhood of e_X.

The group chunks X and Y are said to be *equivalent* if there exist local homomorphisms $f : X \dashrightarrow Y$ and $g : Y \dashrightarrow X$ such that $f \circ g$ is equivalent to Id_Y and $g \circ f$ is equivalent to Id_X.

We make analogous definitions in the analytic case by requiring all spaces to be manifolds and all maps to be morphisms.

Example. Let G be a topological group and let X be an open neighborhood of e with the obvious group chunk structure. X is a group chunk which is equivalent to a topological group.

One may ask whether every group chunk is equivalent to a topological group. The answer is *yes* in the following two cases: a) finite dimensional analytic group chunks, b) metrizable locally compact group chunks (see R. Jacoby, *Annals of Math.*, **66**, 1957). The answer is *no* for analytic group chunks modelled on Banach spaces (see W. van Est and Th. Korthagen, *Proc. Neder. Akad.*, **67**, 1964).

4. Prolongation of subgroup chunks

Let G be a topological group and let X be a subset of G containing e. Then X is said to be a *subgroup chunk* of G if there exists a neighborhood U of e *in* X such that $x, y \in U$ implies $xy \in X$ and $x^{-1} \in X$.

Suppose X is a subgroup chunk of G. Define a subgroup N of G as follows:

$$N = \{ g \in G : \text{for some open neighborhood } U \text{ of } e \text{ in } G, U \cap X = U \cap g^{-1} X g \}.$$

It is clear that N is a subgroup of G and that there is a neighborhood U of e in X such that $U \subset N$. Let $\imath : N \to G$ be the inclusion. We have:

Theorem. *Let* $F = \{ U \cap N : U$ *is a neighborhood of* e *in* $X \}$. *Then:*

1. *F satisfies the axioms for a filter base of neighborhoods of e in N compatible with the group structure in N.*

2. *Suppose N is given the topology defined in F. Then $\imath$ is continuous and gives an equivalence of the group chunks N and X.*

Proof. 1. We verify axioms (GV_{I}'), (GV_{II}'), (GV_{III}') of Bourbaki, *TG.*, III.4.

We may suppose that all neighborhoods U of e in X are contained in N by the remark preceding the theorem. Then what we must show is:

a. Given $U \in F$, there exists $V \in F$ such that $V \cdot V \subset U$.

b. Given $U \in F$, there exists $V \in F$ such that $V^{-1} \subset U$.

c. Given $U \in F$ and $g \in N$, there exists $V \in F$ such that $V \subset gUg^{-1}$.

Now, statement a) and b) are an immediate consequence of the fact that the maps $(x, y) \mapsto xy$ and $x \mapsto x^{-1}$ are continuous in G and hence in X. Statement c) is a consequence of the definition of N.

2. It is clear from the definition of the topology in N that $\imath$ is a local homeomorphism of a neighborhood of e in N onto a neighborhood of e in X. In particular, $\imath$ is continuous as a map $N \to G$ at e, hence, is continuous everywhere. See Bourbaki, *TG.*, III.15.

The theorem shows in particular that *every subgroup chunk is equivalent to a topological group.*

Remark. In general, $\imath$ is not a homeomorphism of N onto $\imath N$. Indeed, in the case $X = \{e\}$, $N = G$ with the discrete topology.

Suppose now that G is an analytic group and that X is an analytic subgroup chunk of G. Then, since N is locally homeomorphic to X at e_N and since X is a submanifold of G, $(N, \imath)$ satisfies (Im) at e_N. Hence, $(N, \imath)$ satisfies (Im) by §2, n°2, Remark 1. We may therefore give N the unique structure of analytic group such that $\imath$ is an analytic group homomorphism and an immersion. In particular, N and X are locally equivalent as analytic group chunks.

Let us examine in more detail the case where $k = \mathbf{R}$ or $\mathbf{C}$. Then N is locally connected so that the connected component H of e_N in N is an open and closed group submanifold of N. We call H the *analytic group generated by X.*

Suppose $\imath(H)$ is closed in G. Then, we contend that $\imath$ is a homeomorphism, so that H is in fact a group submanifold of G. Indeed, $\imath(H)$ is closed in G and is therefore a Baire space. Further, H is locally compact and connected, therefore, a denumerable union of compact sets. Our contention is thus a consequence of:

Lemma 1. *Let A and B be topological groups. Suppose:*

1. *A is locally compact and a denumerable union of compact sets.*

2. *B is a Baire space.*

3. *A map $\imath : A \to B$ is a continuous bijective homomorphism.*

Then, $\imath$ is a homeomorphism.

In turn, Lemma 1 is a consequence of:

Lemma 2. *Suppose:*

1. *A is locally compact topological group which is a denumerable union of compact sets.*

2. *B is a Baire space.*

3. *$\phi : A \times B \to B$ is a continuous transitive operation of the group A on B.*

Then, for any $b \in B$, ϕ induces a homeomorphism of A/N_b onto B where N_b is the stabilizer of b ($N_b = \{\, x \in A : \phi(x, b) = b \,\}$).

Proof. See Bourbaki, *Intégration*, Chap. 7, App. I.

5. Homogeneous spaces and orbits

Let G be an analytic group, H a group submanifold of G, and form the left coset space G/H. Then G/H is the quotient space of G defined by the equivalence relation $R = \{ (x,y) \in G \times G : x^{-1}y \in H \}$.

Theorem 1. *R is a regular equivalence relation so that G/H has a unique manifold structure making $G \xrightarrow{\pi} G/H$ a submersion.*

Proof. By Chap. 3, §12, we must verify that:

1. R is a submanifold of $G \times G$.
2. $\mathrm{pr}_2 : R \to G$ is a submersion.

To show 1), first define $p : G \times G \to G$ by $p(x,y) = x^{-1}y$. Then $R = p^{-1}H$. Hence, by Chap. 3, §11, n°2, F, it suffices to show that p is everywhere a submersion. Let $(x,y) \in G \times G$. Let $\phi : G \to G \times G$ be defined by $\phi(z) = (x, xz)$. Then $\phi(x^{-1}y) = (x,y)$ and $p\phi = \mathrm{Id}_G$. Hence p is a submersion at (x,y) by Chap. 3, §10, n°2.

To show 2), consider the composition $G \times H \xrightarrow{\psi} R \xrightarrow{\mathrm{pr}_2} G$ where $\psi(x,h) = (xh, x)$ for $(x,h) \in G \times H$. Then, $\mathrm{pr}_2\, \psi$ is the projection on G which is a submersion. Since ψ is surjective, pr_2 is a submersion.

Remarks. 1) The natural action of G on G/H is analytic. Indeed, we have the following commutative diagram:

$$\begin{array}{ccc} G \times G & \longrightarrow & G \\ {\scriptstyle \mathrm{Id}_G \times \pi}\big\downarrow & & \big\downarrow{\scriptstyle \pi} \\ G \times G/H & \longrightarrow & G/H \end{array} .$$

The vertical maps are surjective submersions and the top map is analytic. Hence the bottom map is analytic.

2) Suppose H is a normal subgroup of G. Then G/H is an analytic group. To show this, use a diagram similar to that in 1) to verify that multiplication is analytic.

Let G be an analytic group, X an analytic manifold, and $\phi : G \times X \to X$ a morphism. We say that *G acts on X via ϕ* if:

1. For all $x \in X$, $\phi(e,x) = x$.
2. For all $x \in X$ and all $g, h \in G$, $\phi(g, \phi(h,x)) = \phi(gh, x)$.

Suppose that G acts on X via ϕ. Then, we shall often use the notation $\phi(g,x) = gx$. Let us introduce for convenience the following morphisms:

1. For $g \in G$:

$$L_g : G \to G \text{ defined by } h \mapsto gh \,.$$
$$M_g : X \to X \text{ defined by } x \mapsto gx \,.$$

2. For $x \in X$: $\phi_x : G \to X$ defined by $g \mapsto gx$.

Note that L_g and M_g are analytic isomorphisms and that $\phi_x = M_g \circ \phi_x \circ L_{g^{-1}}$. We obtain from this formula for ϕ_x the following *homogeneity principle*:

(HP) Let P be a local property. Then ϕ_x possesses P if and only if ϕ_x possesses P at one point of G.

In particular, ϕ_x is an immersion (submersion, subimmersion) if and only if it is such at a single point.

We shall fix $x_0 \in X$ and let $H = \{\, h \in G : hx_0 = x_0 \,\}$ = stabilizer of x_0. Also, we let $\phi_0 = \phi_{x_0}$.

Theorem 2. *Suppose ϕ_0 is a subimmersion. Then:*

1. *H is a group submanifold of G.*

2. *The induced map $\bar{\phi}_0 : G/H \to X$ is an immersion.*

Proof. 1. This is a consequence of the definition of a subimmersion: Chap. 3, §10, n°4.

2. Let $g \in G$. Then $\operatorname{Ker} T_g\phi_0 = T_g(gH)$. Hence $T_{\pi g}\bar{\phi}_0$ is injective. Hence $\bar{\phi}_0$ is an immersion.

Corollary. *Let $\psi : G_1 \to G_2$ be a homomorphism of analytic groups which is a subimmersion. Let $K = \operatorname{Ker}\psi$. Then:*

1. *K is a normal group submanifold of G.*

2. *The induced analytic group homomorphism $\psi : G/K \to G_2$ is an immersion.*

Theorem 3. *Suppose* $\operatorname{char} k = 0$. *Then ϕ_0 is a subimmersion.*

Proof. Let $g_0 \in G$ be such that the rank n of $T\phi_0$ at g_0 is maximal. Then the rank of $T\phi$ equals n in a neighborhood U_0 of g_0. Let P_g be the following property of a point g in G:

(P_g) There exists a neighborhood U of g such that $\operatorname{rank} T\phi_0 = n$ in U.

Then P_g is a local property and P_g is true for $g = g_0$. By the homogeneity principle P_g is valid for all $g \in G$. Thus, ϕ has constant rank and is a subimmersion since $\operatorname{char} k = 0$: Chap. 3, §10, n°4, Th.

Theorem 4. *Suppose that G is locally compact and a denumerable union of compact sets and that $\phi(G) = Gx_0$ is locally closed in X. Then:*

1. *The induced map* $\bar{\phi}_0 : G/H \to Gx_0$ *is a homeomorphism.*

2. *Suppose* ϕ_0 *is a subimmersion. Then* Gx_0 *is a submanifold of* X *and* $\bar{\phi}_0$ *is an isomorphism of manifolds.*

Proof. Apply Lemma 2 of §4.

Corollary. *Suppose* $\operatorname{char} k = 0$. *Then* Gx_0 *is a submanifold of* X *if and only if* Gx_0 *is locally closed in* X.

We shall now study the case of principal G bundles. We shall assume that:

1. For all $x \in X$, ϕ_x is an injective immersion.

2. We are given an analytic map $\psi : X \to B$ where B is an analytic manifold such that, for all $x \in X$, $Gx = \psi^{-1}\psi(x)$, and such that ψ is surjective. We let $R = \{ (y, x) \in X \times X : y = gx \text{ for some } g \in G \}$. Then R is an equivalence relation and ψ induces a bijection $\bar{\psi} : X/R \to B$ which is continuous.

In the system (X, ϕ, G, ψ, B), we shall speak of X as the *total space*, G as the *fibre*, and B as the *base*. Par abus de notation, we shall sometimes write X for the entire system (X, ϕ, G, ψ, B).

Theorem 5. *The following conditions on* (X, ϕ, G, ψ, B) *are equivalent:*

1. ψ *is a submersion.*

2. R *is a regular equivalence relation and* $\bar{\psi}$ *is an analytic isomorphism.*

3. *For all* $b \in B$, *there is a neighborhood* U_b *of* b *in* B *and an analytic map* $\sigma_b : U_b \to \psi^{-1}U_b$ *such that* $\psi \circ \sigma_b = \mathrm{Id}_{U_b}$.

4. *For all* $b \in B$, *there is a neighborhood* U_b *of* b *in* B *and an analytic isomorphism* $\theta_b : G \times U_b \to \psi^{-1}U_b$ *such that:*
   a. *The following diagram is commutative:*

$$\begin{array}{ccc} G \times U_b & \xrightarrow{\theta_b} & \psi^{-1}U_b \\ \mathrm{pr}_2 \downarrow & & \downarrow \psi \\ U_b & \xrightarrow{\mathrm{Id}} & U_b \end{array} .$$

   b. *For* $g, h \in G$ *and* $a \in U_b$, $\theta_b(gh, a) = g\theta_b(h, a)$.

Proof. 1 $\iff$ 2: This is an immediate consequence of Godement's Theorem: Chap. 3, §12.

1 $\Rightarrow$ 3: This is a consequence of the 5th equivalent form of the definition of a submersion: Chap. 3, §10, n°2.

3 $\Rightarrow$ 4: Define $\theta_g : G \times U_b \to \psi^{-1}U_b$ by the formula $\theta_b(g, a) = g \cdot \sigma_b(a)$. Then θ_b is a bijective morphism which satisfies 4a) and 4b). To show that θ_b is an isomorphism, we must show that θ_b is étale at all $(g, a) \in G \times U_b$. Let $x = \theta_b(g, a) = g \cdot \sigma_b(a)$ and let $\sigma = M_g \circ \sigma_b = g \circ \sigma_b$. Then since $\psi \circ \sigma = \mathrm{Id}_{U_b}$,

ψ is a submersion at x. In addition, $T_a\sigma$ is injective and T_xX is the direct sum of $\operatorname{Im} T_a\phi$ and $\operatorname{Ker} T_x\psi$. But, since $\psi^{-1}a = Gx$ and since ϕ_x is an immersion, $\operatorname{Ker} T_x\psi = T_x(Gx) = \operatorname{Im} T_e\phi_x$. Finally, however,

$$T_{g,a}\theta_b = T_e\phi_x \times T_a\sigma \ .$$

Hence θ_g is étale at (g, a).

$4 \Rightarrow 1$: Trivial.

Definition. Suppose the conditions of the preceding theorem are satisfied by (X, ϕ, G, ψ, B). Then X is said to be a *principal G-bundle* over the base B.

Remark. We have been considering G as acting on X on the left. Thus, we have defined what is known as a left principal bundle. A similar definition is made when G acts on the right.

Theorem 6. *Let G be an analytic group and H be a group submanifold of G. Let $\pi : G \to G/H$ be the projection of G onto the left coset space of H and let $\phi : G \times H \to G$ be the multiplication map. Then G is a right principal H-bundle over the base G/H.*

Proof. This is a special case of Theorem 5.

6. Formal groups: definition and elementary examples

Let R be a commutative ring with unit and consider the formal power series ring $R[[X_1, \ldots, X_n]] = R[[X]]$ in n variables. Let $Y = (Y_1, \ldots, Y_n)$ be a second set of n variables.

Definition. A *formal group law* in n variables is an n-tuple $F = (F_i)$ of formal power series, $F_i \in R[[X, Y]]$, such that:

1. $F(X, 0) = X$ and $F(0, Y) = Y$.
2. $F(U, F(V, W)) = F(F(U, V), W)$.

Let us give some examples:

1) Additive group: $F_i(X, Y) = X_i + Y_i$.

2) Multiplicative group ($n = 1$): $F(X, Y) = X + Y + XY$. Note that we obtain this group law by translating the ordinary multiplicative group law from 1 to 0: $(1 + X)(1 + Y) = 1 + X + Y + XY$.

3) Special Case of Witt Groups for a prime p and $n = 2$:

$$F_1(X_1, X_2, Y_1, Y_2) = X_1 + Y_1$$

$$F_2(X_1, X_2, Y_1, Y_2) = X_2 + Y_2 + \frac{1}{p}(X_1^p + Y_1^p - (X_1 + Y_1)^p) \ .$$

We next give some elementary properties of formal groups:

1) Each F_1 has the form:

$$F_i(X,Y) = X_i + Y_i + \sum_{\substack{|\alpha|\geq 1 \\ |\beta|\geq 1}} c_{\alpha,\beta} X^\alpha Y^\beta .$$

This is an immediate consequence of Axiom 1 of a formal group.

2) There exists a unique $\phi(X) = (\phi_1(X), \ldots, \phi_n(X))$ with $\phi_i(X) \in R[[X]]$, such that $\phi(0) = 0$ and:

$$F(X, \phi(X)) = 0 = F(\phi(X), X) .$$

The existence of a unique $\phi(X)$ such that $\phi(0) = 0$ and such that the first equation is satisfies is a consequence of $D^2F(0) = \mathrm{Id}_{R^n}$. See Bourbaki, *A.*, IV.35. The second equation can likewise be solved uniquely by some $\psi(X)$ such that $\psi(0) = 0$. Then

$$\begin{aligned}\psi(X) &= F(\psi(X), 0) = F(\psi(X), F(X, \phi(X))) = F(F(\psi(X), X), \phi(X)) \\ &= F(0, \phi(X)) = \phi(X) .\end{aligned}$$

Remark. Let us indicate how formal groups will be of interest to us. There are two cases of importance:

1. $R = k$, where k is a complete field.
2. $R = A$, where A is the valuation ring of a complete ultrametric field.

In case 1, we shall define a natural functor:

$$\text{Analytic Groups} \xrightarrow{T} \text{Lie Algebras}$$

We shall want to define a functor S in the opposite direction such that $T \circ S = \mathrm{Id}$. The problem of constructing S is just the problem of constructing an analytic group having a given Lie algebra. It will be useful to know that, over a field of characteristic zero, there is an equivalence of categories:

$$\text{Lie Algebras} \longleftrightarrow \text{Formal Groups}$$

The study of case 2 will be a useful tool when we want to study analytic groups over a complete ultrametric field k. We shall have a commutative diagram of functors:

$$\begin{array}{ccc} \text{Analytic Groups}/k & \longrightarrow & \text{Formal Groups}/k \\ & \nwarrow \ \nearrow & \\ & \text{Formal Groups}/A & \end{array}$$

What we will have in this case is that every analytic group is locally just a formal group$/A$.

7. Formal groups: formulae

We shall use the notation $O(d^0 \geq n)$ to stand for a formal power series whose homogeneous parts vanish in degree strictly less than n. We will let $F(X,Y)$ denote a formal group law over a ring R except as otherwise stated.

1) $F(X,Y) = X + Y + B(X,Y) + O(d^0 \geq 3)$, where B is a bilinear form. This is an immediate consequence of the basis expression for a formal group law since the coefficients $c_{\alpha,\beta}$ vanish unless $|\alpha|$ and $|\beta| \geq 1$.

We shall set $[X,Y] = B(X,Y) - B(Y,X)$.

2) Let $\phi(X)$ be the formal inverse corresponding to F. Then:

$$\phi(X) = -X + B(X,X) + O(d^0 \geq 3) .$$

Indeed, write $\phi(X) = \phi_2(X) + \cdots$, where $\phi_i(X)$ is homogeneous of degree i. Then:

$$0 = F(X, \phi(X)) = X + \phi_1(X) + O(d^0 \geq 2) .$$

Hence, $\phi_1(X) = -X$. Using this result, we find:

$$\begin{aligned} 0 &= F(X, \phi(X)) = X + (-X + \phi_2(X) + \cdots) + B(X, -X + \cdots) + \cdots \\ &= \phi_2(X) - B(X,X) + O(d^0 \geq 3) . \end{aligned}$$

Hence, $\phi_2(X) = B(X,X)$.

3) $XYX^{-1} = Y + [X,Y] + O(d^0 \geq 3)$.
Indeed:

$$\begin{aligned} XYX^{-1} &= (X + Y + B(X,Y) + \cdots) \\ &\quad + (-X + B(X,X) + \cdots) \\ &\quad + B(X + Y + \cdots, -X + \cdots) \\ &\quad + \cdots \\ &= Y + [X,Y] + O(d^0 \geq 3) . \end{aligned}$$

In this case, it will later be convenient to have a notation for the higher order terms. We set:

$$XYX^{-1} = Y + [X,Y] + \sum d_{\alpha,\beta} X^\alpha Y^\beta ,$$

where the range of α and β in the sum is: $|\alpha| \geq 1$, $|\beta| \geq 1$, $|\alpha| + |\beta| \geq 3$.

4) $Y^{-1}XY = X + [X,Y] + O(d^0 \geq 3)$.
The proof is similar to that of formula 3.

5) $X^{-1}Y^{-1}XY = [X,Y] + O(d^0 \geq 3)$.
Use formula 4 and apply the same technique of proof as in formula 3.

6) Jacobi: $[X,[Y,Z]] + [Y,[Z,X]] + [Z,[X,Y]] = 0$.
We shall apply the identity of P. Hall (See L.A., Chap. 2, §1):

$$(X^Y, (Y,Z))(Y^Z, (Z,X))(Z^X, (X,Y)) = 0 .$$

We contend that:

$$(X^Y,(Y,Z)) = [X,[Y,Z]] + O(d^0 \geq 4)$$
$$(Y^Z,(Z,X)) = [Y,[Z,X]] + O(d^0 \geq 4)$$
$$(Z^X,(X,Y)) = [Z,[X,Y]] + O(d^0 \geq 4) \; .$$

Indeed, it suffices by symmetry to check the first of the three formulae. To do that, we note that:

$$X^Y = X + O(d^0 \geq 2) \qquad \text{(Formula 4)}$$
$$(Y,Z) = [Y,Z] + O(d^0 \geq 3) \qquad \text{(Formula 5)}$$

Hence, applying again formula 5, we find:

$$(X^Y,(Y,Z)) = [X,[Y,Z]] + O(d^0 \geq 4) \; .$$

Finally, examining the P. Hall identity up to order 3 and using the formulae we have just obtain, we arrive at the Jacobi identity.

7) The m-th power map:
Define $f_0(X) = 0$ and $f_{m+1}(X) = F(X, f_m(X))$. Note that these conditions determine f_m for both positive and negative m. Equivalently, $f_{-m} = \phi \circ f_m$ where m denotes a positive integer. By induction, we find that:

$$f_m(X) = mX + O(d^0 \geq 2) \; .$$

More generally, we have:

Theorem (Lazard). *There exist unique power series:*

$$\psi_1(X) = (\psi_1^{(1)}(X), \ldots, \psi_1^{(n)}(X))$$
$$\vdots$$
$$\psi_i(X) = (\psi_i^{(1)}(X), \ldots, \psi_i^{(n)}(X))$$
$$\vdots$$

such that:

1. $\psi_1(X) = X$.
2. *$\psi_i(X)$ is of order $\geq i$.*
3. *For all $m \in \mathbf{Z}$, $f_m(X) = \sum_{i=1}^{\infty} \binom{m}{i} \psi_i(X)$.*

Proof. The uniqueness statement is obvious from property 3 applied to $m = 1, 2, \ldots$. To prove existence, we reformulate the theorem. Suppose $F(X,Y)$ is a system of n formal power series such that:

a. $F(X,Y) = X + Y + O(d^0 \geq 2)$.

b. $F(0,Y) = Y$.

Define f_m (for $m \in \mathbf{Z}$) by $f_0 = 0$ and $f_{m+1}(X) = F(X, f_m(X))$. Write

$$f_m(X) = \sum_{\alpha} a_\alpha(m) X^\alpha ,$$

where a_α is a map of $\mathbf{Z}$ into $R \times \cdots \times R$ (n factors). We contend that, for each α, a_α is a "binomial polynomial function of degree $\leq |\alpha|$", that is, that there exist elements $a^i_\alpha \in R \times \cdots \times R$ such that:

$$a_\alpha(m) = \sum_{i \leq |\alpha|} a^i_\alpha \binom{m}{i} \qquad \text{for all } m \in \mathbf{Z}.$$

Note that the contention proves the theorem since we may take $\psi_i = \sum_{|\alpha| \geq i} a^i_\alpha X^\alpha$.

We prove the contention by induction on $|\alpha|$.

$|\alpha| = 0$:

We have $a_\alpha = 0$, since the f_m's have zero constant term.

$|\alpha| = 1$:

Assume the result for $|\beta| < |\alpha|$. Now, we wish to show that $a_\alpha(m)$ is a binomial polynomial in m of degree at most $|\alpha|$. It is well known that to do this it suffices to show that $(\Delta a_\alpha)(m) = a_\alpha(m+1) - a_\alpha(m)$ is a binomial polynomial in m of degree at most $|\alpha| - 1$. Write:

$$F(X, Y) = X + Y + \sum c_{\gamma\delta} X^\gamma Y^\delta .$$

Then, by the hypotheses on F, the range of γ and δ in the sum is: $|\gamma| \geq 1$ and $|\gamma| + |\delta| \geq 2$. Now:

$$f_{m+1}(X) = X + f_m(X) + \sum c_{\gamma\delta} X^\gamma (f_m(X))^\delta .$$

If $|\alpha| = 1$, $a_\alpha(m+1) = a_\alpha(m)$ and we are done. If $|\alpha| > 1$, we find, by comparison, that $a_\alpha(m+1) = a_\alpha(m) + S_\alpha(m)$, where $S_\alpha(m)$ is the sum of the coefficients of X^α appearing in each of the terms: $c_{\gamma\delta} X^\gamma (f_m(X))^\delta$. Since $|\gamma| \geq 1$, the only terms which contribute to $S_\alpha(m)$ are those for which $|\delta| < |\alpha|$. Look at $(f_m(X))^\delta$ for such δ. Then the coefficients of $X^{\alpha-\gamma}$ in that product has coordinates which are sums of products of the following form: $\prod_\nu b_{i_\nu}(m)$, where $b_{i_\nu}(m)$ is a coordinate of a coefficient in f_m with total degree i_ν. By induction, b_{i_ν} is a binomial polynomial of degree $\leq i_\nu$. But it is easy to see that a product of binomial polynomials is again a binomial polynomial (cf. Exer. 2, for instance); moreover, the inequality $\sum i_\nu = |\alpha| - |\gamma| < |\alpha|$ shows that $\prod_\nu b_{i_\nu}$ has degree $< |\alpha|$. It then follows that $S_\alpha = \Delta a_\alpha$ is a binomial polynomial of degree $< |\alpha|$; hence a_α is a binomial polynomial of degree $\leq |\alpha|$. q.e.d.

Corollary. *Let p be a prime number. Then $f_p \equiv \psi_p \bmod p$. In particular, f_p has order $\geq p \bmod p$.*

8. Formal groups over a complete valuation ring

Let k be a complete ultrametric field, let A be the valuation ring of k, and let $\mathfrak{m}$ be the maximal ideal of A. Let $F(X,Y)$ be a formal group law over A. Let $G = \{(x_1,\dots,x_n) : x_i \in \mathfrak{m}\} = P_0(1,\dots,1)$. We define a multiplication on G by the formula: $xy = F(x,y)$. We contend that G is an analytic group. We must verify:

1. Associative law.

2. Existence of unit element: 0 will be the unit.

3. Existence of an inverse: $\phi(x)$ will be x^{-1}, where ϕ is the unique formal power series such that $F(X,\phi(X)) = 0 = F(\phi(X),X)$.

Each of these statement is a consequence of the corresponding rule for the formal group F as is shown by the following lemma:

Lemma. *Suppose $f \in A[[X_1,\dots,X_p]]$ and $g_i \in A[[Y_1,\dots,Y_q]]$, $1 \le i \le p$, and suppose $g_i(0) = 0$ for all i. Let $h = f(g_1,\dots,g_p) \in A[[Y_1,\dots,Y_q]]$. Then, for $x_1,\dots,x_q \in \mathfrak{m}$, we have*

$$h(x) = f(g_1(x),\dots,g_p(x)) .$$

Proof. See Bourbaki, *Alg. Comm.*, Chap. 3, §4 n°5, Cor. of Prop. 6.

Definition. A group G constructed in the above manner will be called *standard.*

Theorem. *Any analytic group chunk contains an open subgroup which is standard.*

Proof. Let G be an analytic group chunk. By shrinking G and choosing local coordinates, we may assume that G is an open neighborhood of 0 in k^n and that the multiplication in G is given by a power series $F(X,Y)$ such that F converges on the ball of radius $< \varepsilon$. Write $F(X,Y) = X + Y + \sum c_{\alpha,\beta} X^\alpha Y^\beta$. Here, $|\alpha|$ and $|\beta|$ are equal to or greater than 1 and the coefficients $c_{\alpha,\beta}$ are vectors in k^n. We shall change coordinates by multiplication by $\mu \in k$. Specifically, if $x, y \in G$ are such that $z = xy$ is defined, set $x' = \mu x$, $y' = \mu y$, and $z' = \mu z$. Then:

$$z' = x' + y' + \sum \frac{c_{\alpha,\beta}}{\mu^{|\alpha|+|\beta|-1}} x'^{\alpha} y'^{\beta} .$$

Hence, the group law F_μ in the new coordinates has coefficients $\frac{c_{\alpha,\beta}}{\mu^{|\alpha|+|\beta|-1}}$. By choosing μ so that $|\mu|$ is sufficiently large, we can insure that the coefficients of F_μ lie in A^n and that $|\mu|\varepsilon \ge 1$ so that F converges in the ball of radius 1. In the coordinate system defined by μ, the strict unit ball is a standard subgroup of G.

Corollary 1. *Any analytic group chunk is equivalent to an analytic group.*

Corollary 2. *Any analytic group chunk has a basis of neighborhoods of* e *consisting of open subgroups.*

9. Filtrations on standard groups

The notation and assumptions of §8 will be used throughout this section. In addition, we let $w : k \to \mathbf{R} \cup \{\infty\}$ be the valuation of k, that is, for some $a \in \mathbf{R}$, $0 < a < 1$, we have, for all $x \in k$:

$$|x| = a^{w(x)} .$$

For $x = (x_1, \ldots, x_n) \in k^n$, we let $w(x) = \inf_i(w(x_i))$. For $\lambda \geq 0$, we let:

$$G_\lambda = \{\, x \in G : w(x) \geq \lambda \,\}$$
$$G_\lambda^+ = \{\, x \in G : w(x) > \lambda \,\} .$$

More generally, for an ideal $\mathfrak{a}$ of A, we let:

$$G_\mathfrak{a} = \{\, x \in G : x_i \in \mathfrak{a} \text{ for } 1 \leq i \leq n \,\}$$
$$G_\mathfrak{a}^+ = \{\, x \in G : x_i \in \mathfrak{a} \cdot \mathfrak{m} \text{ for } 1 \leq i \leq n \,\}$$

Thus, if $\mathfrak{a}_\lambda = \{\, x \in A : w(x) \geq \lambda \,\}$, we have $G_\lambda = G_{\mathfrak{a}_\lambda}$ and $G_\lambda^+ = G_{\mathfrak{a}_\lambda}^+ = G_{\mathfrak{a}_\lambda \cdot \mathfrak{m}}$.

Theorem 1. *For all ideals $\mathfrak{a}$ of A, $G_\mathfrak{a}$ and $G_\mathfrak{a}^+$ are normal subgroups of G. Moreover, if $x, y \in G$, the relation $x \equiv y \bmod G_\mathfrak{a}$ is equivalent to $x_i \equiv y_i \bmod \mathfrak{a}$ for $1 \leq i \leq n$.*

Proof. Let $G(A/\mathfrak{a})$ be the group of systems $\bar{x} = (\bar{x}_1, \ldots, \bar{x}_n)$ where $\bar{x}_i \in \mathfrak{m}/\mathfrak{a}$, the multiplication being given by the reduction $\bar{F}$ of F mod $\mathfrak{a}$. If $x \in G$, the reduction $\bar{x}$ of x mod $\mathfrak{a}$ is defined and $x \mapsto \bar{x}$ is a surjective homomorphism $\phi_\mathfrak{a} : G \to G(A/\mathfrak{a})$. The kernel of $\phi_\mathfrak{a}$ is $G_\mathfrak{a}$; this proves Theorem 1 for $G_\mathfrak{a}$. The assertions about $G_\mathfrak{a}^+$ follow since $G_\mathfrak{a}^+ = G_{\mathfrak{a}\cdot\mathfrak{m}}$.

(Alternate proof: use formula 1) of §6 and formulae 2), 3) of §7.)

Corollary. *The subsets $\{G_\lambda\}$ define a filtration of G.*

Proof. We must verify the axioms of a filtration (cf. L.A., Chap. 2, §2):

1. $w(0) = \infty$.
2. For all $x \in G$, $w(x) > 0$.
3. For all $x, y \in G$, $w(xy^{-1}) \geq \inf\{w(x), w(y)\}$.
4. For all $x, y \in G$, $w((x, y)) \geq w(x) + w(y)$.

Axioms 1 and 2 are obvious from the definition of G. Axiom 3 is equivalent to the assertion that G_λ is a subgroup of G for all λ. Axiom 4 is equivalent to the assertion that $(G_\lambda, G_\mu) \subset G_{\lambda+\mu}$. In fact, if $x \in G_\lambda$ and $y \in G_\mu$, we have

1) $[x, y] \in G_{\lambda+\mu}$.

2) $(x, y) \equiv [x, y] \pmod{G^+_{\lambda+\mu}}$.

Now 1) is clear and 2) follows from Theorem 1 and formula 5) of §7.

Theorem 2. *Let $\mathfrak{a}$ and $\mathfrak{b}$ be ideals of A such that $\mathfrak{a} \subset \mathfrak{b} \subset \mathfrak{a}^2$. The reduction map $\phi_{\mathfrak{b}} : G \to G(A/\mathfrak{b})$ induces an isomorphism of the group $G_{\mathfrak{a}}/G_{\mathfrak{b}}$ onto the additive group $(\mathfrak{a}/\mathfrak{b})^n$.*

Proof. Formula 1) of §6 shows that, if $x, y \in G_{\mathfrak{a}}$, $F(x, y) \equiv x + y \bmod \mathfrak{a}^2$. The theorem follows from this and from Theorem 1.

Corollary 1. *Let $\lambda \in w(\mathfrak{m})$, $\lambda \neq \infty$. Then G_λ/G^+_λ is isomorphic to the additive group $(A/\mathfrak{m})^n$.*

Proof. Choose $a \in \mathfrak{m}$ such that $w(a) = \lambda$ and let $\mathfrak{a} = (a)$. Then, by Theorem 2, $G_{\mathfrak{a}}/G^+_{\mathfrak{a}}$ is isomorphic to $(\mathfrak{a}/\mathfrak{a}\mathfrak{m})^n$. However, the map: $\alpha \mapsto \alpha a$ defines an isomorphism of $(A/\mathfrak{m})$ onto $(\mathfrak{a}/\mathfrak{a}\mathfrak{m})$ which proves the corollary.

Corollary 2. *Suppose that k is locally compact and that $p = \mathrm{char}(A/\mathfrak{m})$. Then:*

1. *For all $\lambda \in w(\mathfrak{m})$, $\lambda \neq \infty$, G_λ/G^+_λ is a commutative finite p-group.*
2. *For all $\lambda \in w(\mathfrak{m})$, $\lambda \neq \infty$, G_λ/G^+_λ is a p-group.*
3. *G is a projective limit of p-groups ("pro p-group").*

Proof. We note that since k is locally compact:

1) $A/\mathfrak{m}$ is compact and discrete, hence finite.

2) $\mathfrak{m}$ is compact so that w takes on a minimum value on some element $a \in \mathfrak{m}$.

Then $\mathfrak{m} = (a)$ so that A is a discrete valuation ring.

Statement 1) is then a consequence of Corollary 1 and 1) above. Statement 2) is a consequence of statement 1) and 2) above. Statement 3) is a consequence of statement 2).

We shall now use the filtration $\{G_\lambda\}$ of G to study the r-th power maps f_r (cf. §7). We let $\bar{k} = A/\mathfrak{m}$ be the residue field of k and let $= \mathrm{char}\,\bar{k}$.

Theorem 3. *Suppose r is relatively prime to p. Then, for all $\lambda \in w(\mathfrak{m})$, $\lambda \neq \infty$, f_r defines an analytic manifold isomorphism of G_λ onto G_λ.*

Proof. The image of r in $\bar{k}$ is a unit in $\bar{k}$ so that r is a unit in A. Hence, f_r is an invertible formal power series in $[[A]]$. Let $\theta = f_r^{-1}$. Then θ is absolutely convergent on G and $f_r \circ \theta = \theta \circ f_r = \mathrm{Id}$ by the lemma quoted in §8. Since f_r and θ preserve G_λ, f_r is a bijection on G_λ. Finally, the derivative of f_r at each $x \in G$ is congruent mod $\mathfrak{m}$ to $r \cdot \mathrm{Id}$ and hence is invertible. Thus, f_r is étale and hence is an analytic isomorphism on G_λ.

Theorem 4. *Suppose* $\operatorname{char} k = 0$ *and that* $p \neq 0$. *Let* $\mu = w(p)$. *Then, for all* $\lambda \in w(\mathfrak{m})$, $\frac{\mu}{p-1} < \lambda < \infty$, f_p *is an analytic manifold isomorphism of* G_λ *onto* $G_{\lambda+\mu}$.

Proof. By Lazard's Theorem, $f_p(X) = p(X + \phi(X)) + \psi(X)$, where $\operatorname{ord}\phi \geq 2$ and $\operatorname{ord}\psi \geq p$. Now, for $x \in G_\lambda$ and for α with $|\alpha| \geq 1$, we have that $w(x^\alpha) \geq \lambda|\alpha|$. In particular:

1) $w(\phi(x)) > \lambda$.

2) $w(\psi(x)) \geq p\lambda \geq \lambda + (p-1)\lambda > \lambda + \nu$.

It follows that $f_p(G_\lambda) \subset G_{\lambda+\mu}$. To show that f_p is an analytic isomorphism of G_λ onto $G_{\lambda+\mu}$, we choose $a \in G$ such that $w(a) = \lambda$ and consider the function $F : A^n \to A^n$ defined by $F(X) = \frac{1}{ap} f_p(ax)$. Then:

$$F(X) = X + \frac{1}{a}\phi(aX) + \frac{1}{ap}\psi(aX) .$$

Let $r \in \mathbf{R}$, $0 < r < 1$, be such that $|a|$, $\frac{|a|^{p-1}}{|p|} < r^{p-1}$. Then:

1) The coefficients of degree $i \geq 2$ in $\frac{1}{a}\phi(aX)$ have absolute value less than or equal to $|a|^{i-1} \leq r^{i-1}$.

2) The coefficients of degree $i \geq p$ in $\frac{1}{ap}\psi(aX)$ have absolute value less than or equal to $|a|^{i-p}\frac{|a|^{p-1}}{|p|} \leq r^{i-1}$.

We shall see in Appendix 2 that these conditions imply that F and its formal inverse θ converge absolutely on A^n. In particular, we may actually compose F and θ on A^n. This shows that F is an analytic isomorphism of A^n onto A^n. It is then immediate that $f_p : G_\lambda \to G_{\lambda+\mu}$ is an analytic isomorphism.

Theorem 5. *Let* G *be an analytic group over* k. *Then there exists an open subgroup* U *which contains no finite subgroup* H *such that* $\operatorname{ord} H$ *is prime to* $\operatorname{char} k$.

Proof. Since G contains an open subgroup which is standard, the theorem is reduced to Theorem 3 and 4.

Remark. In particular, when $\operatorname{char} k = 0$, Theorem 5 asserts that G contains no "small" finite subgroups.

We shall give some applications of Theorem 5 in Appendix 3.

Exercises

1. Let k be locally compact, and let A be a compact analytic group over k.
   a) Let G be a finite group of order prime to the characteristic of k. Assume G acts analytically on A. Define as usual the 1st-cohomology set $H^1(G,A)$ (resp. the higher cohomology groups $H^q(G,A)$ if A is abelian). Prove that $H^1(G,A)$ is finite (Hint: use the manifold structure of the cocycles). Prove analogous results for $H^q(G,A)$, $q \geq 1$, when A is abelian.
   b) Using a), prove that the finite subgroups of A of given order (prime to char k) are in finite number, up to conjugation.
2. Let i, j be two positive integers.
   a) Prove *a priori* that $\binom{m}{i}\binom{m}{j}$, as a function of m, is equal to a linear combination of binomials $\binom{m}{k}$, with $i,j \leq k \leq i+j$.
   b) Prove the identity:
$$\binom{m}{i}\binom{m}{j} = \sum_{i,j\leq k\leq i+j} \frac{k!}{(k-i)!(k-j)!(i+j-k)!}\binom{m}{k}.$$
   (Hint: compute in two ways the series $(1+X)^m(1+Y)^m$, where X and Y are indeterminates.)
3. Notations being those of §7, 7) (Lazard's theorem) consider the case of an $F(X,Y)$ with property a ($F(X,Y) \equiv X+Y \bmod \deg 2$), but not property b ($F(0,Y) = Y$). Show that it is still possible to write the f_m's in the form $\sum \binom{m}{i}\psi_i$, but that it is not true in general that $\operatorname{ord}(\psi_i) \geq i$.
4. Show that Lemma 2 of §4 remains true if hypothesis (1) is replaced by: (1′) - A is a complete Hausdorff topological group (for both uniform structures), and its topology can be defined by a denumerable family of open sets. (Hint: imitate the proof of Banach's closed graph theorem.)
5. Let k be a locally compact ultrametric field, and let G be a standard group of dimension n over k. Let dx be a Haar measure on the additive group k^n. Show that the restrictions of dx to G (which is open in k^n) is a left and right Haar measure on G. (Hint: use the fact that $G = \varprojlim G/G_\lambda$, and that a Haar measure on G is an inverse limit of Haar measures on the finite groups G/G_λ).
6. a) Let $F(X,Y) = X + Y + XY$ be the "multiplicative" formal group law in one variable. Show that the ψ_i's of Lazard's theorem are just the monomials X^i.
   b) Assume moreover that k is ultrametric, of characteristic zero and residue characteristic p. Show that the following are equivalent:

   1) $f_p(X) = 0$

   2) $1 + x$ is a p-th root of unity in k.

Using the theorem 4 of §9, show that this implies $w(x) \geq x(p)/(p-1)$. Show that it is in fact an equality if $x \neq 0$ (i.e., if $1+x$ is a primitive p-th root of unity).

7. Let F and F' be two group laws over a field k of characteristic p and let $x' = \phi(x)$ be a formal homomorphism of F into F' (i.e., $\phi(F(x,y)) = F'(\phi(x),\phi(y))$). Assume the terms of degree one in ϕ are all zero. Show that ϕ is a power series in x^p. (Hint: use the differential equation

$$\phi'(x) \cdot D_2F(x,0) = D_2F'(\phi(x),0) \cdot \phi'(0)$$

to show that $\phi' = 0$.) Interpret this result as a factorization of ϕ through a "Frobenius map" $F \to F^{(p)}$, when k is perfect.

Appendix 1. Maximal compact subgroups of GL(n, k)

We prove here the theorem stated in §2, n°1.

Let k be a locally compact ultrametric field, A be the valuation ring of k, $\mathfrak{m}$ be the maximal ideal of A, and $G = \mathrm{GL}(n, A)$ for some $n > 0$, $n \in \mathbf{Z}$.

Lemma 1. *Let L be an A-submodule of k^n. Then, the following are equivalent:*

1. *L is finitely generated over A and L generates k^n over k.*
2. *L is free of rank n over A.*

Proof. 1 ⇒ 2: Since A is a principal ideal domain, L is free; $\operatorname{rank}_A L = n$, since L generates a k^n over A.

2 ⇒ 1: Trivial:

An A-submodule L of k^n satisfying the equivalent conditions of Lemma 1 is called a *lattice* in k^n.

Lemma 2. *Let $L_1, \ldots, L_r$ be lattices in k^n and let L be the A-submodule of k^n generated by $L_1, \ldots, L_r$. Then L is a lattice in k^n.*

Proof. We verify 1) of Lemma 1. Clearly L generates k^n over k since each L_i does. Since, moreover, each L_i is finitely generated over A, the module L which they generate over A is finitely generated over A.

Lemma 3. *Let L be a lattice in k^n and let K_L be the subgroup of* GL(n, k) *which send L onto L. Then, for some $\alpha \in \mathrm{GL}(n,k)$, $K_L = \alpha G \alpha^{-1}$. In particular, K_L is compact and open.*

Proof. By 2) of Lemma 1, we may choose $\alpha \in \mathrm{GL}(n,k)$ such that $\alpha(A^n) = L$. Then, by the definition of $\mathrm{GL}(n,A) = G$, $K_L = \alpha \cdot G \cdot \alpha^{-1}$. We have already noted that G is compact and open; hence, K_L is compact and open.

Lemma 4. *Let L and L' be two lattices in k^n and suppose $K_L \subset K_{L'}$. Then there exists $\lambda \in k^*$ such that $L' = \lambda \cdot L$, and $K_L = K_{L'}$.*

Proof. It is clear that $K_{L'}$ does not change when L' is replaced by $\lambda \cdot L'$, with $\lambda \in k^*$. We can then suppose that $L' \subset L$ and $L' \not\subset \mathfrak{m} \cdot L$. Let $V = L/\mathfrak{m}L$ and $V' = (L' + \mathfrak{m}L)/\mathfrak{m}L$; V' is a non-zero vector subspace of V (over the residue field $A/\mathfrak{m}$). Moreover, since $K_L \subset K_{L'}$, the lattice L' is invariant by K_L, hence its image V' in V is invariant by K_L, i.e, by $\mathrm{GL}(V)$. Since $V' \neq 0$, this implies $V' = V$, hence $L' + \mathfrak{m}L = L$ and, by a standard argument (Nakayama's lemma!) $L' = L$. q.e.d.

Theorem 1. *Let H be a compact subgroup of* $\mathrm{GL}(n,k)$. *Then:*

1. *There exists a lattice M in k^n such that H sends M onto M.*

2. *There exists $\alpha \in \mathrm{GL}(n,k)$ such that $H \subset \alpha \cdot G \cdot \alpha^{-1}$.*

Proof. 1. Choose any lattice L in k^n, for example, $L = A^n$. Then $H_L = H \cap K_L$ is exactly the subgroup of H which sends L onto L. Since K_L is open in $\mathrm{GL}(n,k)$, H_L is open in H. Hence H_L has finite index in H since H/H_L is compact and discrete, therefore, finite. Therefore, the number of translates σL of L, $\sigma \in H$, is finite. Let M be the A-submodule generated by $\{\sigma L\}_{\sigma \in H}$. It is clear that H sends M onto M and it follows from Lemma 2 that M is a lattice.

2. This statement follows immediately from 1) and Lemma 3.

Theorem 2. 1. *G is a maximal compact subgroup of* $\mathrm{GL}(n,k)$.

2. *The maximal compact subgroups of* $\mathrm{GL}(n,k)$ *are precisely the conjugates of G.*

3. *Every compact subgroup of* $\mathrm{GL}(n,k)$ *is contained in a maximal compact subgroup of* $\mathrm{GL}(n,k)$.

Proof. 1. Suppose G is contained in a compact subgroup H of $\mathrm{GL}(n,k)$. Theorem 1 shows that there exists a lattice M such that $H \subset K_M$. Hence $G \subset K_M$, and, by Lemma 4, $G = K_M$. Hence G is maximal.

Assertions 2 and 3 follow from 1 and Theorem 1.

Appendix 2. Some convergence lemmas

Suppose that $F(X) = (F_i(X))$ is a system of n formal power series in n variables and suppose that each $F_i(X)$ has the form:

$$F_i(X) = X_i - \sum_{|\alpha| \geq 2} a^i_\alpha X^\alpha = X_i - \phi_i(X) .$$

We have seen in the proof on the Inverse Function Theorem that the system F is formally invertible and that we may write the formal inverse system $\theta(X) = (\theta_i(X))$ where each $\theta_i(X)$ has the form:

$$\theta_i(X) = X_i + \sum_{|\beta|\geq 2} b^i_\beta X^\beta = X_i + \psi_i(X) .$$

Suppose $r \in \mathbf{R}$ and $0 < r < 1$. Consider the conditions:

(A_r) For all α, $|a^i_\alpha| \leq r^{|\alpha|-1}$.

(B_r) For all β, $|b^i_\beta| \leq r^{|\beta|-1}$.

Lemma 1. $(A_r) \Rightarrow F$ *converges absolutely on* A^n.
$(B_r) \Rightarrow \theta$ *converges absolutely on* A^n.

Proof. It suffices to remark that:

$$\sum_{|\gamma|\geq 0} r^{|\gamma|} = \frac{1}{(1-r)^n} < \infty .$$

Lemma 2. $(A_r) \iff (B_r)$.

Proof. It suffices by symmetry to show that $(A_r) \Rightarrow (B_r)$. We show this by induction on $|\beta|$, that is, we assume the statement true for β', $|\beta'| < |\beta|$, and we prove it for β. Now:

$$X_i = F_i(\theta(X)) = \theta_i(X) - \phi_i(\theta(X)) .$$

Comparing the coefficients of X^β, we find that b^i_β is the sum of the coefficients of X^β in $\phi_i(\theta(X))$. Since k is ultrametric, it suffices to show that each time X^β occurs in $\phi_i(\theta(X))$ its coefficient satisfies the estimate desired for b^i_β. Now:

$$\phi_i(\theta(X)) = \sum_{|\alpha|\geq 2} a^i_\alpha(\theta(X))^\alpha$$

and

$$\theta(X)^\alpha = \theta_1(X)^{\alpha_1}\cdots\theta_n(X)^{\alpha_n} .$$

A typical monomial term in $\theta(X)^\alpha$ has the form:

$$\prod_{i=1}^{n}\prod_{j=1}^{\alpha_i}(b^i_{\gamma_{i,j}}X^{\gamma_{i,j}}) .$$

We are interested in terms where $\sum \gamma_{i,j} = \beta$. Then, we can estimate the product of all the coefficients in that product by:

$$\prod_{i,j} r^{|\gamma_{i,j}|-1} = r^{|\beta|-|\alpha|} .$$

Since $|a^i_\alpha| \le r^{|\alpha|-1}$, we obtain the desired final estimate of $r^{|\beta|-1}$ for the coefficients of X^β in $\phi_i(\theta(X))$.

Corollary. *$(A_r) \Rightarrow F$ is an analytic isomorphism of A_n onto A_n.*

Proof. By Lemma 2, we have both (A_r) and (B_r). Then, from Lemma 1, for $x \in A^n$, $x = (f \circ \theta)(x) = f(\theta(x)) = (\theta \circ f)(x) = \theta(f(x))$.

Appendix 3. Applications of §9: "Filtrations on standard groups"

Theorem 1. *For each $n > 0$, $n \in \mathbf{Z}$, there exists $N > 0$, $N \in \mathbf{Z}$, such that any finite subgroup of* $\mathrm{GL}(n, \mathbf{Q})$ *has order* $\le N$.

Proof. 1. We prove first the corresponding statement for the p-adic integers $\mathbf{Z}_p$, for any prime p. By Theorem 5 of §9, there exists an open normal subgroup U of $\mathrm{GL}(n, \mathbf{Z}_p)$ such that U contains no non-trivial finite subgroups. Then, if $H \subset \mathrm{GL}(n, \mathbf{Z}_p)$ is a finite group, $H \subset \mathrm{GL}(n, \mathbf{Z}_p)/U$ and $\mathrm{ord}\, H \le N$ where $N = \mathrm{ord}\, \mathrm{GL}(n, \mathbf{Z}_p)/U$.

2. We reduce the theorem to the statement we have proved in 1). We use two different methods:

Method 1: Let $H \subset \mathrm{GL}(n, \mathbf{Q})$ be finite. Let p be a prime and consider $H \subset \mathrm{GL}(n, \mathbf{Q}_p)$. Then, H is compact, so some conjugate of H is contained in $\mathrm{GL}(n, \mathbf{Z}_p)$ by Theorem 1 of Appendix 1. Hence, $\mathrm{ord}\, H \le N$, where N is the bound of 1).

Method 2: Note that Lemmas 1 and 2 of Appendix 2 are valid for $k = \mathbf{Q}$ and $A = \mathbf{Z}$. We have, in addition, the following statements:

1. Let L be a lattice in $\mathbf{Q}_n$. Then the subgroup of $\mathrm{GL}(n, \mathbf{Q})$ which sends L onto L is a conjugate of $\mathrm{GL}(n, \mathbf{Z})$.

2. If H is a finite subgroup of $\mathrm{GL}(n, \mathbf{Q})$, there exists a lattice M which H sends onto itself.

We prove statement 1) in exactly the way the corresponding statement in Lemma 3 of Appendix 1 is proved. To prove statement 2), let L be any lattice and define M to be the lattice generated by the finite set of lattices: $\{\sigma L\}_{\sigma \in H}$. Then H sends M onto M as desired.

Combining statements 1) and 2), we see that if H is a finite subgroup of $\mathrm{GL}(n, \mathbf{Q})$, then H has a conjugate in $\mathrm{GL}(n, \mathbf{Z})$, hence to prove it suffices to consider finite subgroups of $\mathrm{GL}(n, \mathbf{Z})$. Now, $\mathrm{GL}(n, \mathbf{Z}) \subset \mathrm{GL}(n, \mathbf{Z}_p)$ so we may again reduce the theorem to what we have already shown.

We may obtain explicit estimates for the integer N in Theorem 1 by taking the gcd of the estimates at each prime p. Consider:

1. p odd:
   Then $1 > \frac{1}{p-1}$ so we may take

$$U = G_1 = \{ y : y = 1 + x,\ x = (x_{ij}),\ x_{ij} \in \mathfrak{m} \} .$$

Then, $\mathrm{GL}(n, \mathbf{Z}_p)/U = \mathrm{GL}(n, \mathbf{F}_p)$ where $\mathbf{F}_p$ is the field with p elements. We can compute the order of $\mathrm{GL}(n, \mathbf{F}_p)$ explicitly: it is simply the number of distinct sets of ordered bases in $\mathbf{F}_p^n$. This number is:

$$(p^n - 1)(p^n - p) \cdots (p^n - p^{n-1}) .$$

2. $p = 2$:
   Then $2 > \frac{1}{p-1}$ so we may take $U = G_2$. Then $\mathrm{GL}(n, \mathbf{Z}_p)/U = \mathrm{GL}(n, \mathbf{Z}/4\mathbf{Z})$. We have an exact sequence:

$$0 \longrightarrow (2\mathbf{Z}/4\mathbf{Z})^{n^2} \longrightarrow \mathrm{GL}(n, \mathbf{Z}/4\mathbf{Z}) \longrightarrow \mathrm{GL}(n, \mathbf{Z}/2\mathbf{Z}) \longrightarrow 1 .$$

Hence, the number of elements in $\mathrm{GL}(n, \mathbf{Z}/4\mathbf{Z})$ is:

$$2^{n^2}(2^{n-1})(2^n - 2) \cdots (2^n - 2^{n-1}) .$$

Let us look more closely at the case $n = 2$:

1. p odd:
   The number computed above becomes: $(p^2 - 1)(p^2 - p) = (p-1)^2 p(p+1)$. Now, for p odd, we have
   a. $p^2 - 1 \equiv 0 \pmod 8$, hence $(p^2 - 1)(p^2 - p) \equiv 0 \pmod{16}$.
   b. $(p - 1)p(p + 1) \equiv 0 \pmod 3$.
   Hence, the number computed above is congruent to 0 mod 48. When $p = 3$, we have that $(p^2 - 1)(p^2 - p) = 48$.

2. $p = 2$:
   The number is $2^{2^2}(2^2 - 1)(2^2 - 2) = 96$.

Hence, by the above method, the best estimate we obtain for the order of finite subgroups of $\mathrm{GL}(2, \mathbf{Q})$ is 48. The situation is in fact somewhat better. We first note that any finite subgroup of $\mathrm{GL}(2, \mathbf{Q})$ is contained in the set of matrices of determinant $\pm 1 = G_0$. We have an exact sequence:

$$1 \longrightarrow \mathrm{SL}(2, \mathbf{Q}) \longrightarrow G_0 \longrightarrow \mathbf{Z}/2\mathbf{Z} \longrightarrow 0 .$$

Hence to obtain an estimate for the order of the finite subgroups of G_0, we need only multiply the corresponding estimate for $\mathrm{SL}(2, \mathbf{Q})$ by 2. We shall show that

1. Every finite subgroup of $\mathrm{SL}(2, \mathbf{Q})$ is a subgroup of a rotation group on the plane and is therefore cyclic.

Only cyclic group of order 1, 2, 3, 4, 6 occur in $\mathrm{SL}(2, \mathbf{Z})$.

Proof. 1. Let $H \subset \mathrm{SL}(2, \mathbf{Q})$ be finite and let B be any positive definite bilinear form on $\mathbf{Q}^2$. Let $\overline{B}(x,y) = \sum_{\sigma \in H} B(\sigma x, \sigma y)$. Then, $\overline{B}$ is positive definite and H leaves $\overline{B}$ invariant. Since the elements of H have determinant 1, H is a subset of the rotations of $\mathbf{R}^2$ with respect to the scalar product $\overline{B}$.

2. Let σ be an element of finite order in $\mathrm{SL}(n, \mathbf{Q})$. We pass to $\mathbf{C}$ and put α in Jordan canonical form:

Case 1: α has the form:

$$\alpha = \begin{pmatrix} \mu & 1 \\ 0 & \mu \end{pmatrix} .$$

Then , an easy calculation shows that α does not have finite order which is a contradiction.

Case 2: α has the form:

$$\alpha = \begin{pmatrix} \mu & 0 \\ 0 & \nu \end{pmatrix} .$$

Let $N = \mathrm{ord}(\alpha)$. Then $\mu^N = \nu^N = 1$. Hence μ and ν are roots of unity. We also know that μ and ν lie in a quadratic extension of $\mathbf{Q}$ since they satisfy the characteristic polynomial of α. In fact if either μ or ν is not in $\mathbf{Q}$, both of them are not in $\mathbf{Q}$ and they are complex conjugates. Hence, we can have only the following cases:

a. $\mu = \nu = 1$ or $\mu = \nu = -1$.

b. μ is a primitive N^{th}-root of unity, $N > 2$, and $\nu = \bar{\mu}$. Since the Nth-cyclotomic field is of degree $\phi(N)$ over $\mathbf{Q}$ (ϕ being Euler's function), we have $\phi(N) = 2$, hence $N = 3$, 4, or 6. This proves the second statement.

Let us give explicitly elements of order 4 and order 6 in $\mathrm{SL}(2, \mathbf{Z})$. In each case we shall find the appropriate matrix by considering a quadratic extension $K = \mathbf{Q}(x)$ of $\mathbf{Q}$ and representing multiplication by x using the basis $\{1, x\}$ of K over $\mathbf{Q}$.

1. An element of order 4:

Take x to be a primitive 4-th root of unity. Then multiplication by x has order 4 and is represented by the matrix:

$$x = \begin{pmatrix} 0 & -1 \\ 1 & 0 \end{pmatrix} .$$

2. An element of order 6:

Take x to be a primitive 6-th root of unity. Then multiplication by x has order 6 and is represented by the matrix:

$$x = \begin{pmatrix} 0 & -1 \\ 1 & 1 \end{pmatrix} .$$

* * *

Let k be a locally compact ultrametric field, A be the valuation ring of k, $\mathfrak{m}$ be the maximal ideal of A, $p = \operatorname{char} k$, and $q = \operatorname{Card}(A/\mathfrak{m})$. Let w be the canonical valuation of k, that is, $w : k \to \mathbf{Z} \cup \{\infty\}$ with $w(k^*) = \mathbf{Z}$. Then the canonical absolute value on k is defined equivalently as follows:

1. For $x \in A$, $\|x\| = \operatorname{Card}(A/xA)^{-1}$.
2. For $x \in k$, $\|x\| = q^{-w(x)}$.
3. Multiplication by x alters the Haar measure on k by $\|x\|$.

Suppose r is relatively prime to p. Consider $f_r : A^* \to A^*$. Let

$$s = \operatorname{Card}(\operatorname{Ker} f_r) = \text{number of roots of unity in } k \text{ with exponent dividing } r\ .$$

Theorem 2. $\operatorname{Card}(A^*/A^{*r}) = \|r\|^{-1} \cdot s$.

We shall obtain Theorem 2 as a consequence of a more general theorem on analytic groups over k. So let G be a *commutative compact analytic group* over k and define:

$$h_r(G) = \operatorname{Card}(\operatorname{Coker} f_r)/\operatorname{Card}(\operatorname{Ker} f_r)\ .$$

We shall see in Theorem 3 that $h_r(G)$ is well defined, that is, that both numbers on the right hand side are finite, and we shall compute $h_r(G)$. We shall let $n = \dim_k G$.

Theorem 3. *The number $h_r(G)$ is well defined and equal to $\|r\|^{-n}$.*

Proof. We shall prove the following three statements which imply the theorem:

1. The theorem is true of $G = H_\lambda$, where H is a standard group and $\lambda \gg 0$.

2. The theorem is true if G is a finite group.

3. The theorem is true for G if G contains a normal subgroup manifold H such that the theorem is true for H and G/H.

We shall prove the statements in reverse order:

3. Consider the commutative diagram with exact rows:

$$\begin{array}{ccccccccc} 1 & \longrightarrow & H & \longrightarrow & G & \longrightarrow & G/H & \longrightarrow & 1 \\ & & \phi_1\downarrow & & \phi_2\downarrow & & \phi_3\downarrow & & \\ 1 & \longrightarrow & H & \longrightarrow & G & \longrightarrow & G/H & \longrightarrow & 1 \end{array}$$

where $\phi_1 = f_r$ on H, $\phi_2 = f_r$ on G, and $\phi_3 = f_r$ on G/H. Then, there exists an exact sequence (Bourbaki, *Alg. Comm.*, Chap. 1, §1, n°4):
(*)

$$1 \to \operatorname{Ker}\phi_1 \to \operatorname{Ker}\phi_2 \to \operatorname{Ker}\phi_3 \xrightarrow{\delta} \operatorname{Coker}\phi_1 \to \operatorname{Coker}\phi_2 \to \operatorname{Coker}\phi_3 \to 1$$

Aside from δ, all the maps are defined in the obvious manner. We define δ as follows: let $x'' \in \operatorname{Ker}\phi_3$ and choose $x \in G$ such that x maps to x'' mod H; then, $x'' \in \operatorname{Ker}\phi_3$ implies $x^r \in H$; define $\delta(x'')$ = image of x^r in $\operatorname{Coker}\phi_1$. It is left to the reader to verify that δ is well defined and that $(*)$ is exact.

Since the theorem is assumed true for H and G/H, we have that $\operatorname{Ker}\phi_1$, $\operatorname{Ker}\phi_3$, $\operatorname{Coker}\phi_1$, and $\operatorname{Coker}\phi_3$ are finite. Therefore, $\operatorname{Ker}\phi_2$ and $\operatorname{Coker}\phi_2$ are finite since $(*)$ is exact. Abbreviate Card by c. Then, the exactness of $(*)$ implies in addition that:

$$1 = c(\operatorname{Ker}\phi_1)c(\operatorname{Ker}\phi_2)^{-1}c(\operatorname{Ker}\phi_3)c(\operatorname{Coker}\phi_1)^{-1}c(\operatorname{Coker}\phi_2)c(\operatorname{Coker}\phi_3)^{-1} .$$

In other words:

$$h_r(G) = h_r(H) \cdot h_r(G/H) .$$

Finally to obtain the explicit formula for $h_r(G)$, let $m = \dim_k H$. Then, we have also that: $n - m = \dim_k G/H$. Then:

$$\|r\|^{-n} = \|r\|^{-m}\|r\|^{-(n-m)} .$$

Since by hypothesis $h_r(H) = \|r\|^{-m}$ and $h_r(G/H) = \|r\|^{-(n-m)}$, we obtain by comparing the above two formulae that $h_r(G) = \|r\|^{-n}$, as desired.

2. We have an exact sequence:

$$1 \longrightarrow \operatorname{Ker} f_r \longrightarrow G \longrightarrow G \longrightarrow \operatorname{Coker} f_r \longrightarrow 1 .$$

Then, since G is finite, $\operatorname{Ker} f_r$ and $\operatorname{Coker} f_r$ are finite, $n = 0$, and:

a. $1 = c(\operatorname{Coker} f_r)c(G)^{-1}c(G)c(\operatorname{Ker} f_r)^{-1} = h_r(G)$.

b. $1 = \|r\|^{-n}$.

This proves 2.

1. We need only prove this part for large $\lambda \in \mathbf{Z}$. Hence, by Theorem 3 and 4 of §9, we may assume that λ lies in the range such that $f_r : G_\lambda \to G_{\lambda+w(r)}$ is an isomorphism. Then:

a. $c(\operatorname{Ker} f_r) = 1$.

b. $c(\operatorname{Coker} f_r) = (q^{w(r)})^n = \|r\|^{-n}$.

This proves 1 and the theorem as well.

Exercise. Using, for example, Haar measure, one may show that if $G \xrightarrow{\phi} G$, where ϕ is an analytic étale group endomorphism of G, then

1. $\operatorname{Ker}\phi$ and $\operatorname{Coker}\phi$ are finite.

2. $h_\phi = c(\operatorname{Coker}\phi)/c(\operatorname{Ker}\phi) = \|\det T_e\phi\|^{-1}$.

Chapter V. Lie Theory

Unless otherwise specified, k will denote a field complete with respect to a non-trivial absolute value.

1. The Lie algebra of an analytic group chunk

Suppose $F(X,Y)$ is a formal group law over k. Then we have seen that:

1. $F(X,Y) = X + Y + B(X,Y) + O(d^0 \geq 3)$, where $B(X,Y)$ is a bilinear form on k^n (Chap. 4, §7, n°1).

2. Define $[X,Y]_F = B(X,Y) - B(Y,X)$. Then $[X,Y]_F$ defines on k^n a Lie algebra structure (Chap. 4, §7, n°6).

We say that $[X,Y]_F$ is the Lie algebra associated to the formal group F.

Now suppose that G is an analytic group chunk over k. Let $L(G) = \mathfrak{g} = T_eG$. We define a canonical Lie algebra structure on $\mathfrak{g}$ as follows. Choose a chart $c = (U, \phi, n)$ of G at e. Then the group law on G is induced via ϕ from a formal group law F on k^n. Let $\bar{\phi} : \mathfrak{g} \to k^n$ be the isomorphism which is determined by ϕ. Then, for $x, y \in \mathfrak{g}$, define:

$$\bar{\phi}[x,y]_c = [\bar{\phi}x, \bar{\phi}y]_F \ .$$

We contend that $[x,y]_c$ is in fact independent of the choice of c. To show this we prove the following lemma:

Lemma 1. *Let G and G' be analytic group chunks, c and c' charts at e and e', and $f : G \longrightarrow G'$ a local homomorphism. Then $T_ef : \mathfrak{g} \to \mathfrak{g}'$ is a Lie algebra homomorphism with respect to the structures $[\ ,\ ]_c$ and $[\ ,\ ]_{c'}$.*

Proof. The proof is immediately reduced to:

Lemma 2. *Let $F(X,Y)$ and $F'(X',Y')$ be two formal group laws and let f be a formal homomorphism from F and F'. Let f_1 be the linear part of f. Then:*

$$[f_1(X), f_1(Y)]_{F'} = f_1([X,Y]_F) \ .$$

Proof. From Chap. 4, §7, n°5, we have:

$$\begin{aligned} f(X)^{-1}f(Y)^{-1}f(X)f(Y) &= [f_1(X), f_1(Y)]_{F'} + O(d^0 \geq 3) \\ f(X^{-1}Y^{-1}XY) &= f_1([X,Y]_F) \quad + O(d^0 \geq 3) \end{aligned} \ .$$

Comparing the terms of degree 2, we obtain the lemma.

Definition. In the above setting, we say that $\mathfrak{g}$ together with its canonical Lie algebra structure is the *Lie algebra of G*.

Remark. Lemma 1 shows that the construction of the Lie algebra of a group chunk is functorial.

2. Elementary examples and properties

1) The Lie algebra of a general linear group.

Suppose R is an associative algebra with unit which is finite dimensional over k. We have seen (Chap. 4, §2, n°1) that $G_m(R)$ is an analytic group which is an open subset of R. Hence $T_1 G_m(R) = R$. Multiplication in $G_m(R)$ has the form:

$$(1+x)(1+y) = 1 + x + y + xy \ .$$

This law of multiplication corresponds to the formal group law:

$$F(x,y) = x + y + xy \ .$$

Hence, the Lie algebra structure on $T_1 G_m(R) = R$ is given by:

$$[x,y] = xy - yx \ .$$

In particular, we obtain the usual Lie algebra structure when R is the endomorphism ring $E(V)$ of a finite dimensional vector space V.

2) The Lie algebra of a product

Suppose G_1 and G_2 are analytic groups. Then the linear isomorphism of $T_e(G_1 \times G_2)$ with $T_{e_1}G_1 \times T_{e_2}G_2$ is a Lie algebra isomorphism. Indeed, let c_i be a chart at e_i on G_i and let $c = c_1 \times c_2$ be the product of the charts c_i. Then, the product of the Lie algebras $[\ ,\ ]_{c_i}$ is $[\ ,\ ]_c$.

3) The Lie algebra of a group submanifold

Suppose G and H are analytic groups and that $f : H \to G$ is an analytic group homomorphism which is an immersion. Then $L(f) : L(H) \to L(G)$ is injective. Hence $L(H)$ is identified with a Lie subalgebra of $L(G)$. In particular, we may apply this remark when H is a group submanifold of G and f is the inclusion.

Let us consider in more detail the case where $\operatorname{char} k = 0$.

Theorem. *Suppose* $\operatorname{char} k = 0$ *and let* H_1 *and* H_2 *be group submanifolds of* G. *Then* $H_1 \cap H_2$ *is a group submanifold of* G *and* $L(H_1 \cap H_2) = L(H_1) \cap L(H_2)$.

Proof. By Theorem 1 of Chap. 4, §5, G/H_1 is a manifold. Let x be the coset H_1 in G/H_1. Then H_2 acts on G/H_1 and the stabilizer of x is $H_1 \cap H_2$. Since $\operatorname{char} k = 0$, $H_1 \cap H_2$ is a group submanifold of G (Chap. 4, §5, Thms. 2 and 3). Finally, $L(H_1 \cap H_2)$ may be identified with its image in $L(G)$ which is the kernel of the map: $T_e H_2 \to T_e G / T_e H_1$, that is, with $L(H_1) \cap L(H_2)$.

Corollary 1. *Suppose* $L(H_1) \subset L(H_2)$. *Then, in a neighborhood of* e, $H_1 \subset H_2$.

Proof. Indeed, $T_e(H_1 \cap H_2) = L(H_1) \cap L(H_2) = L(H_1) = T_e(H_1)$, and $H_1 \cap H_2 \subset H_1$. Hence, $H_1 \cap H_2$ and H_1 agree in a neighborhood of e, that is, $H_1 \subset H_2$ in a neighborhood of e.

Corollary 2. *Suppose $L(H_1) = L(H_2)$. Then, in a neighborhood of e, $H_1 = H_2$.*

Corollary 3. *Let G_1 and G_2 be analytic groups and let $\phi, \psi : G_1 \to G_2$ be analytic group homomorphisms. Then, ϕ and ψ agree in a neighborhood of e_1 if and only if $L(\phi) = L(\psi)$.*

Proof. The graphs G_ϕ and G_ψ of ϕ and ψ in $G_1 \quad G_2$ are group submanifolds. Now using the identification of n°2, we have:

$$L(G_\phi) = \{ (x,y) \in L(G_1 \times G_2) : y = L(\phi)(x) \}$$
$$L(G_\psi) = \{ (x,y) \in L(G_1 \times G_2) : y = L(\psi)(x) \} .$$

Hence, by Corollary 2, the following statements are equivalent:

a. ϕ and ψ agree in a neighborhood of e_1.

b. G_ϕ and G_ψ agree in a neighborhood of (e_1, e_2).

c. $L(\phi) = L(\psi)$.

This proves the corollary.

4) The Lie algebra of a kernel

Suppose G and H are analytic groups and $\phi : G \to H$ is an analytic group homomorphism which is a subimmersion. Let $K = \operatorname{Ker} \phi$. Then, we have seen, in Chap. 4, §5, Cor. of Thm. 2, that K is a group submanifold of G. Moreover, we have:

$$L(K) = \operatorname{Ker} T_e\phi = \{ x \in L(G) : L(\phi)(x) = 0 \} .$$

3. Linear representations

Let G be an analytic group and V be a vector space. Then, a *linear representation* of G in V is an analytic group homomorphism $\sigma : G \to \mathrm{GL}(V)$. The group G acts on V via σ:

$$g \cdot v = \sigma(g)(v) .$$

We obtain from σ an induced representation of $L(G)$ via the induced homomorphism $\bar{\sigma} : L(G) \to E(V)$ of Lie algebras.

1) Basic examples

1. The identity representation: $\mathrm{GL}(V) \to \mathrm{GL}(V)$.

2. Let V^* denote the dual of V. Define $^* : \mathrm{GL}(V) \to \mathrm{GL}(V^*)$ to be the map $u \mapsto {}^t u^{-1}$. Then, * is an analytic group isomorphism. Let $1 = \mathrm{Id}_V$ and $1^* = \mathrm{Id}_{V^*}$. We have in a neighborhood of 1:

$$*(1+x) = (1^* + {}^t x)^{-1} = \sum_{\mu=0}^{\infty} (-1)^\mu ({}^t x)^\mu = 1^* - {}^t x + O(d^0 \geq 2) .$$

In particular:

$$L(*)(x) = -{}^t x .$$

3. Let $V_1, \ldots, V_n$ be vector spaces and set $V = V_1 \otimes \cdots \otimes V_n$. Define:

$$\theta : E(V_1) \times \cdots \times E(V_n) \to E(V) ,$$

by

$$\theta(u_1, \ldots, u_n) = u_1 \otimes \cdots \otimes u_n .$$

Then, θ induces an analytic group homomorphism of $\prod_{i=1}^n \mathrm{GL}(V_i)$ to $\mathrm{GL}(V)$. In a neighborhood of 1 in $G = \prod_{i=1}^n \mathrm{GL}(V_i)$, we have:

$$\theta(1+x_1, \ldots, 1+x_n) = 1 + \sum_{i=1}^n 1 \otimes \cdots \otimes x_i \otimes \cdots \otimes 1 + O(d^0 \geq 2) .$$

In particular:

$$L\theta(x_1, \ldots, x_n) = \sum_{i=1}^n 1 \otimes \cdots \otimes x_i \otimes \cdots \otimes 1 .$$

4. Let $V_1, \ldots, V_n$ and W be vector spaces and set $V = L(V_1, \ldots, V_n; W)$ and $G = (\prod_{i=1}^n \mathrm{GL}(V_i)) \times \mathrm{GL}(W)$. Then, V is canonically isomorphic to $V_1^* \otimes \cdots \otimes V_n^* \otimes W$. We may therefore apply Examples 2 and 3 to obtain a map $\theta : G \to \mathrm{GL}(V)$. This map is given explicitly by:

$$\theta(u_1, \ldots, u_n, w)(v) = w \circ v \circ (u_1 \otimes \cdots \otimes u_n)^{-1} .$$

Translating our previous result, we find that:

$$L\theta(x_1, \ldots, x_n, z)(y) = z \circ y - \sum_{i=1}^n y \circ (1 \otimes \cdots \otimes x_i \otimes \cdots \otimes 1) .$$

5. Let $G = \mathrm{GL}(V)$ and consider the analytic homomorphism

$$\det : G \to G_m(k) ,$$

where det denotes the determinant map and we have viewed k^* as $G_m(k)$. In a neighborhood of 1 in G, we have:

$$\det(1+x) = 1 + \mathrm{tr}(x) + \cdots + \det(x) = 1 + \mathrm{tr}(x) + O(d^0 \geq 2) ,$$

where tr denotes the trace map. In particular:

$$L(\det)(x) = \mathrm{tr}(x) .$$

2) Kernels of representations

We may apply example 4) of §2 whenever a linear representation is a subimmersion. In particular, let us apply it to *5* of n°1. The determinant map is a submersion (if $V \neq 0$) so that the hypothesis is satisfied. We define:

$$\mathrm{SL}(V) = \text{special linear group} = \mathrm{Ker}(\det) .$$

In particular, combining the calculations of §1, n°4, and n°, *5*, above, we obtain:

$$L(\mathrm{SL}(V)) = \{\, x \in E(V) : \mathrm{tr}(x) = 0 \,\} .$$

3) Stabilizer subgroups

Assume char $k = 0$. Then we have seen that if G acts on X and $x \in X$, the stabilizer $G_x = \{\, g \in G : g \cdot x = x \,\}$ of x is a group submanifold of G (Chap. 4, §5, Thms. 2 and 3). We now apply this to representations.

1. Let $\sigma : G \to \mathrm{GL}(V)$ be a linear representation of an analytic group G in a vector space V and let $v \in V$. Considering the action of G on V induced by σ, we have that G_v is a group submanifold of G and:

$$L(G_v) = \{\, x \in L(G) : \bar{\sigma}(x)(v) = 0 \,\} .$$

Indeed, let $\phi : \mathrm{GL}(V) \to V$ be the map $u \mapsto u(v)$. Then $T_e\phi : E(V) \to V$ is the map $y \mapsto y(v)$. Since $T_e G_v = \mathrm{Ker}(T_e(\phi\sigma)) = \mathrm{Ker}(T_e\phi \circ \bar{\sigma})$, the desired result is proved.

2. With notation as in *1*, let $f \in V^*$ and consider the stabilizer subgroup G_f with respect to the representation $* \circ \sigma : G \to \mathrm{GL}(V^*)$. Then:

$$G_f = \{\, g \in G : f \circ \sigma(g) = f \,\}$$
$$L(G_f) = \{\, x \in L(G) : f \circ \bar{\sigma}(x) = 0 \,\} .$$

To prove the first statement, it suffices, since G_f is a group, to show that $g^{-1} \in G_f$ if and only if $f \circ \sigma(g) = f$. But:

$$* \circ \sigma(g^{-1})(f) = {}^t\sigma(g^{-1})^{-1}(f) = f \circ \sigma(g) .$$

The result is then a consequence of the definition of G_f. To prove the second statement, it suffices from *1* to show that $L(* \circ \sigma)(x)(f) = 0$ is equivalent to $f \circ \bar{\sigma}(x) = 0$. But, by n°s 1, 2, we have:

$$L(* \circ \sigma)(x)(f) = -{}^t\bar{\sigma}(x)(f) = -f \circ \bar{\sigma}(x) .$$

The desired equivalence is clear.

A particular kind of group which may be obtained in the above manner is the affine group $A(V)$ of a vector space V. We identify V with the group of translations on V. Then, $A(V)$ is the semi-direct product of V and $\mathrm{GL}(V)$. The group law is given by:

$$(v_1, g_1)(v_2, g_2) = (v_1 + g_1 v_2, g_1 g_2) .$$

We may identify $A(V)$ with the subgroup G of $\mathrm{GL}(V \times k)$ of transformations which leave hyperplanes parallel to V invariant. We do this using a map $\sigma : A(V) \to G$ which is defined as follows:

$$\sigma(v,g)(w,\alpha) = (\alpha v + gw, \alpha) .$$

The group G is however just the group $\mathrm{GL}(V \times k)_f$ with respect to the linear form $f : V \times k \to k$ defined by $f(w,\alpha) = \alpha$.

3. Let $\sigma : G \to \mathrm{GL}(V)$ be a linear representation of an analytic group G in a vector space V and let $\beta \in (V \otimes V)^*$. Let $\theta : \mathrm{GL}(V)^2 \to \mathrm{GL}(V \otimes V)$ be the analytic homomorphism defined in n°1, *3*, when $V = V_1 = V_2$. Consider the composite representation $\tau = \theta \circ (\sigma \times \sigma)$ of G into $\mathrm{GL}(V \otimes V)$. Then, applying *2*, we obtain:

$$G_\beta = \{ g \in G : \beta(\sigma(g) \times \sigma(g)) = \beta \}$$
$$L(G_\beta) = \{ x \in L(G) : \beta(\bar\sigma(x) \otimes 1 + 1 \otimes \bar\sigma(x)) = 0 \} .$$

The condition defining $L(G_\beta)$ is equivalent to:

$$\beta(\bar\sigma(x)v, w) + \beta(v, \bar\sigma(x)w) = 0 , \qquad \text{for all } v, w \in V.$$

There are two applications of the preceding discussion to the case when $G = \mathrm{GL}(V)$, σ is the identity representation, and $V = k^n$, which are of particular interest:

A) Orthogonal Group

Take β to be the bilinear form on k^n defined by:

$$\beta(x_1, \ldots, x_n; y_1, \ldots, y_n) = \sum_{i=1}^{n} x_i y_i .$$

The group G_β is called the orthogonal group of k^n. Let us determine G_β and $L(G_\beta)$ explicitly. First note that for $u \in E(k^n)$:

$$\beta(ux, y) = \beta(x, {}^t u y) , \qquad \text{for all } x, y \in k^n.$$

Thus, we find that for $x, y \in k^n$:

a. For $g \in \mathrm{GL}(k^n)$, $\beta(gx, gy) = \beta(x,y) \iff \beta(x, {}^t g g y) = \beta(x,y)$

b. For $u \in E(k^n)$, $\beta(ux, y) + \beta(x, uy) = 0 \iff \beta(x, ({}^t u + u)y) = 0.$

Since β is non-degenerate, we obtain:

$$G_\beta = \{ g \in \mathrm{GL}(k^n) : {}^t g g = 1 \}$$
$$L(G_\beta) = \{ u \in E(k^n) : {}^t u + u = 0 \} .$$

B) Symplectic Group

Take $n = 2m$ and take β to be the bilinear form defined by:

$$\beta(x_1, \ldots, x_{2m}; y_1, \ldots, y_{2m}) = \sum_{i=1}^{\infty} (x_i y_{m+i} - x_{m+i} y_i) .$$

The group G_β is called the symplectic group of k^{2m}. Let us determine G_β and $L(G_\beta)$ explicitly. We identify $E(k^m \times k^m)$ and $E(k^n)$. Thus, we may write a linear map $u : k^n \to k^n$ in the form:

$$u = \begin{pmatrix} A & B \\ C & D \end{pmatrix} ,$$

where $A, B, C, D \in E(k^m)$. Given a matrix u as above, define u' by:

$$u' = \begin{pmatrix} {}^tD & -{}^tB \\ -{}^tC & {}^tA \end{pmatrix} .$$

Then, one checks that for $x, y \in k^n$:

$$\beta(ux, y) = \beta(x, u'y) .$$

Using this fact and the fact that β is non-degenerate, it then is easy to show that:

$$\begin{aligned} G_\beta &= \{\, g \in \mathrm{GL}(k^n) : g'g = 1 \,\} \\ L(G_\beta) &= \{\, u \in E(k^n) : u' + u = 0 \,\} \end{aligned} .$$

The conditions defining $L(G_\beta)$ can be shown to be equivalent to the three conditions ${}^tA + D = 0$; ${}^tB = B$; ${}^tC = C$.

4. The method of stabilizer subgroups can be applied whenever one combines several representations using the basic representation of n°1. We leave the formulation of the general statement to the reader and instead present one additional example of its use. Let A be a finite dimensional algebra over k and let $\beta : A \otimes A \to A$ be the multiplication map. Let $G = \mathrm{GL}(A)$. Then, the following statements are equivalent, for $g \in A$:

a. $g \in G_\beta$.

b. For all $x, y \in A$: $g\beta(g^{-1}x, g^{-1}y) = \beta(x, y)$.

c. For all $x, y \in A$: $g(xy) = (gx)(gy)$.

Indeed, a) and b) are equivalent by definition; the equivalence of b) and c) follows from replacing x by gx and y by gy in b) to obtain c). Hence, G_β is just the group of automorphisms of the algebra A. We shall now show that $L(G_\beta)$ is the space of derivations of A into A. Indeed, the statement that $d \in L(G_\beta)$ amounts to:

For all $x, y \in A$: $$d\beta(x, y) - \beta(dx, y) - \beta(x, dy) = 0 .$$

The formula in the above condition is simply: $d(xy) = (dx)y + x(dy)$, which proves the contention.

4) Adjoint representation

Let G be an analytic group and consider, for each $g \in G$, the inner automorphism $\phi_g : G \to G$ defined by $x \mapsto gxg^{-1}$. Let $\mathfrak{g} = L(G)$ and let $\mathrm{Ad} : G \to \mathrm{GL}(\mathfrak{g})$ be the map $g \mapsto T_e\phi_g$. Clearly Ad is a group homomorphism and we shall see in a moment that Ad is analytic. Hence, we can consider $L(\mathrm{Ad}) : \mathfrak{g} \to E(\mathfrak{g})$. We shall show that this map is just the map $\mathrm{ad} : \mathfrak{g} \to E(\mathfrak{g})$ which may be defined in general for Lie algebras.

Since Ad is a group homomorphism, we need only check that Ad is analytic at e to show that Ad is analytic everywhere. We shall compute Ad in a neighborhood of e in local coordinates. By Chap. 4, §7, n°3, we have:

$$\phi_g(x) = x + [g, x] + \sum d_{\alpha,\beta} g^\alpha x^\beta ,$$

where $|\alpha| \geq 1$, $|\beta| \geq 1$, $|\alpha| + |\beta| \geq 3$. Hence, $\mathrm{Ad}(g) = T_e \phi_g$ is the map:

$$x \mapsto x + [g, x] + \sum_{|\beta|=1} d_{\alpha,\beta} g^\alpha x^\beta .$$

This shows that Ad is analytic at e. Moreover, since $|\beta| = 1$ in the last sum, $|\alpha| \geq 2$ in the last sum. Hence, $T_e \,\mathrm{Ad}(y)$ is the map $: x \mapsto [y, x]$. Thus, $L(\mathrm{Ad})(y) = T_e \,\mathrm{Ad}(y) = \mathrm{ad}(y)$, as desired.

4. The convergence of the Campbell-Hausdorff formula

Theorem 1. *Let* $\mathrm{char}\, k = 0$ *and let* $\mathfrak{g}$ *be a finite dimensional Lie algebra over* k. *Then* $\mathfrak{g}$ *is the Lie algebra of an analytic group chunk.*

Proof. Let $n = \dim \mathfrak{g}$. We shall make use of the Campbell-Hausdorff formula (L.A., Chap. 4, §§7 and 8) to define a formal group law in n variables satisfying:

1. F is convergent
2. $\mathfrak{g}$ is isomorphic to the Lie algebra k^n under $[\ ,\ ]_F$.

We divide the proof into several steps:

1) Let $x_1, \ldots, x_n$ be a basis of $\mathfrak{g}$. Then, there exists unique "structure constants" $\gamma_{ij}^h = \gamma_{ij}^h(x_1, \ldots, x_n)$ such that, for all i and j:

$$[x_i, x_j] = \sum_{h=1}^{n} \gamma_{ij}^h x_h .$$

Define $\gamma = \gamma(x_1, \ldots, x_n) = \max |\gamma_{ij}^h|$. We observe how γ_{ij}^h and γ behave under the change of basis defined by multiplication by $\lambda \neq 0, \in k$:

$$\gamma_{ij}^h(\lambda x_1, \ldots, \lambda x_n) = \lambda \gamma_{ij}^h(x_1, \ldots, x_n)$$
$$\gamma(\lambda x_1, \ldots, \lambda x_n) = |\lambda| \gamma(x_1, \ldots, x_n) .$$

2) Let $R = k[[X, Y]] = k[[X_1, \ldots, X_n, Y_1, \ldots, Y_n]]$ and let $E = R^n$. Corresponding to the basis $x_1, \ldots, x_n$ of $\mathfrak{g}$, define a Lie algebra structure on E by the formula:

$$[(f_i), (g_j)] = \left(\sum_{i=1}^{n} \sum_{j=1}^{n} \gamma_{ij}^h f_i g_j \right) .$$

In particular, let us consider $\mathrm{ad}(X)$ and $\mathrm{ad}(Y)$ where $X = (X_1, \ldots, X_n)$ and $Y = (Y_1, \ldots, Y_n)$.

We say that $P = (P_1, \ldots, P_n) \in E$ is a homogeneous polynomial of degree r if each P_i is a homogeneous polynomial of degree r. We let $\|P\|$ be the maximum of the absolute value of coefficients of the P_i in this case.

Lemma 1. *Suppose $P \in E$ is homogeneous of degree r. Then, if $Z = X$ or Y, $\mathrm{ad}(Z)(P)$ is homogeneous of degree $(r+1)$ and $\|\mathrm{ad}(Z)(P)\| \leq n^2\gamma\|P\|$.*

Proof. Consider, for example, $Z = X$. Let $\mathrm{ad}(X)(P) = (Q_1, \ldots, Q_n)$ and let $P_j = \sum a^j_{\alpha,\beta} X^\alpha Y^\beta$ and $Q_h = \sum b^h_{\alpha,\beta} X^\alpha Y^\beta$. Then:

$$Q_h = \sum \gamma^h_{ij} X_i P_j \ .$$

Hence, each Q_h is homogeneous of degree $(r+1)$ and in particular:

$$b^h_{\alpha,\beta} = \sum_{i=1}^{n} \sum_{j=1}^{n} \gamma^h_{ij} a^j_{\alpha-\delta_i,\beta} \ .$$

Thus: $|g^h_{\alpha,\beta}| \leq n^2\gamma\|P\|$.

Let $\mathfrak{m}$ denote the maximal ideal of R. Then:

Corollary. *Let $Z = X$ or Y and let $r \geq 0$. Then $\mathrm{ad}(Z)(\mathfrak{m}^r E) \subset \mathfrak{m}^{r+1} E$.*

3) Let $S = \{\bar{x}, \bar{y}\}$ be a set with two elements and consider the free Lie algebra L_S on S and its completion $\hat{L}_S = \prod_{r=0}^{\infty} L^r_S$ (L.A., Chap. 4, §§ 3 and 7). Let θ be the canonical Lie algebra homomorphism from L_S to E such that $\bar{x} \mapsto X$ and $\bar{y} \mapsto Y$.

Lemma 2. *For $r > 0$, $\theta(L^r_S) \subset \mathfrak{m}^r E$.*

Proof. This follows immediately from the corollary to Lemma 1 of n°2.

In particular, θ extends uniquely to a Lie algebra homomorphism of $\hat{L}_S$ into E.

Now, let $\bar{z} \in \hat{L}_S$ be the unique element such that $e^{\bar{x}} e^{\bar{y}} = e^{\bar{z}}$ (L.A., Chap. 4, §7, Thm. 7.4). We let $F = \theta(\bar{z})$. It follows easily from the remarks at the end of §7 of Chap. 4 of L.A. and from arguments similar to those of Lemma 2 that:

1. F is a formal group law in X and Y.
2. $B(X,Y) = \frac{1}{2}[X,Y]$.

In particular: $[X,Y]_F = \frac{1}{2}[X,Y] - \frac{1}{2}[Y,X] = [X,Y]$.

We now prove:

Theorem 2. *F is convergent.*

Proof. We shall need two elementary lemmas:

Lemma 3. *The following sum is convergent for t sufficiently small:*

$$\sum_{\substack{p_1,\dots,p_m \\ q_1,\dots,q_m \\ p_i+q_i\geq 1 \\ m\geq 1}} t^{\sum p_i+\sum q_i} .$$

Proof. The above formal sum may be written as:

$$\sum_{m=1}^{\infty}\Big(\sum_{p+q\geq 1} t^{p+q}\Big)^m .$$

For $t<1$, the term in parenthesis converges to α where:

$$\alpha = \frac{1}{(1-t)^2} - 1 .$$

For t sufficiently small, $\alpha < 1$. Then, the first sum is just a geometric series with ratio less than 1, hence it converges.

Lemma 4. *There exists a constant a,* $0 < a \leq 1$, *such that:*

$$|n!| \geq a^n \quad \textit{and} \quad |n| \geq a^n \qquad (n \in \mathbf{Z},\ n>0).$$

Proof. We consider three cases:

A) k is archimedian.
Take $a = 1$.

B) k is ultrametric and the restriction of the absolute value to $\mathbf{Q}$ is trivial.
Take $a = 1$.

C) k is ultrametric and the restriction of the absolute value to $\mathbf{Q}$ is some p-adic absolute value.
First note that the second inequality follows from the first since we have: $|(n-1)!| \leq 1$, for all $n \geq 1$. Then take $a = |p|^{1/(p-1)}$. We have:

$$|n!| \geq a^n \iff v_p(n!) \leq \frac{n}{p-1} .$$

However:

$$v_p(n!) = [\frac{n}{p}] + [\frac{n}{p^2}] + \cdots \leq \frac{n}{p} + \frac{n}{p^2} + \cdots = \frac{n}{p-1} .$$

We now prove Theorem 2 which will at the same time complete the proof of Theorem 1.

We note first that since change of coordinates by multiplication by a non-zero constant does not effect convergence we may assume that the basis $x_1, \ldots, x_n$ of $\mathfrak{g}$ has been replaced by a basis $\lambda x_1, \ldots, \lambda x_n$ of $\mathfrak{g}$ so that:

$$|\lambda|\gamma \leq \frac{1}{n^2} .$$

We use the old notation $x_1, \ldots, x_n$ for the new basis $x_1, \ldots, x_n$. We then have from Lemma 1:

Lemma 1′. *Suppose $P \in E$ is homogeneous of degree r. Then, if $Z = X$ or Y, $\mathrm{ad}(Z)(P)$ is homogeneous of degree $(r+1)$ and $\|\mathrm{ad}(Z)(P)\| \leq \|P\|$.*

Next, we use Dynkin's formula (L.A., Chap. 4, §8) to write down explicitly $F = \theta(\bar{z})$. We find that $F(X,Y) = \sum_\nu f_\nu(X,Y)$ where the homogeneous part f_ν of F of degree ν may be written as:

$$f_\nu(X,Y) = \frac{1}{\nu} \sum_{p+q=\nu} (f'_{p,q}(X,Y) + f''_{p,q}(X,Y)) .$$

Here:

$$f'_{p,q} = \sum_{\substack{p_1+\cdots+p_m=p \\ q_1+\cdots q_{m-1}=q-1 \\ p_i+q_i\geq 1 \\ p_m\geq 1}} \frac{(-1)^{m+1}}{m} \frac{\mathrm{ad}(X)^{p_1}\,\mathrm{ad}(Y)^{q_1}\cdots \mathrm{ad}(X)^{p_m}(Y)}{p_1!q_1!\cdots p_m!}$$

$$f''_{p,q} = \sum_{\substack{p_1+\cdots+p_{m-1}=p-1 \\ q_1+\cdots q_{m-1}=q \\ p_i+q_i\geq 1}} \frac{(-1)^{m+1}}{m} \frac{\mathrm{ad}(X)^{p_1}\,\mathrm{ad}(Y)^{q_1}\cdots \mathrm{ad}(Y)^{q_{m-1}}(X)}{p_1!q_1!\cdots q_{m-1}!}$$

By Lemma 1′, each of the numerators in the above expression is a homogeneous polynomial of degree ν whose coefficients have absolute value equal to or less than 1. Also, using the expression for [,] we see by induction that the number of monomials actually appearing in each numerator is equal to or less than $n^{2\nu}$. Hence, we can majorize each numerator by a real polynom of degree ν in X and Y which is the sum of $n^{2\nu}$ monomials each of which has coefficient 1. Such a real polynomial is estimated at a radius vector $(s, \ldots, s) \in \mathbf{R}^{2n}$ by:

$$n^{2\nu}s^\nu = (n^2 s)^{p+q} .$$

On the other hand, by Lemma 3, the integers which appear in the denominator can be estimated by:

$$|\nu| \geq a^\nu = a^{p+q}$$

$$|m| \geq a^m \geq a^{p+q}$$

$$\left.\begin{matrix} |p_1!q_1!\cdots p_m!| \\ |p_1!q_1!\cdots q_{m-1}!| \end{matrix}\right\} \geq a^{\sum p_i + \sum q_i} = a^{p+q} .$$

Hence, setting $t = (n^2 s/a^3)$, we see that the components of the formal group law F can be majorized by twice the formal sum in Lemma 3. Since that sum converges for t sufficiently small, F converges for s sufficiently small as was to be shown.

Remarks. 1) When $\mathfrak{g}$ is a nilpotent Lie algebra (L.A., Chap. 5, §2) then F is a polynomial so that the convergence is trivial.

2) To obtain an estimate for the radius of convergence, one might estimate the constants in the proof, namely:

1. The constant a in Lemma 4.
2. The radius of convergence of the series in Lemma 3.

In fact, it is easy to see that the series in Lemma 3 converges for:

$$t < \frac{\sqrt{2}-1}{\sqrt{2}} .$$

Hence, F converges on every polydisk of radius $(R, \ldots, R)$ where R satisfies:

$$R < \frac{a^3}{n^2} \frac{\sqrt{2}-1}{\sqrt{2}} .$$

This estimate is not particular good and no good estimate is known when $k = \mathbf{R}$ or $\mathbf{C}$.

Suppose that k is ultrametric and the the restriction of the absolute value to $\mathbf{Q}$ is some p-adic absolute value. Suppose that $\mathfrak{g}$ is a Lie subalgebra of $L(G_m(R))$ where R is finite dimensional associative algebra with unit over k, that is, $\mathfrak{g} \subset R$ and $[x, y] = xy - yx \in \mathfrak{g}$ for all $x, y \in \mathfrak{g}$. Assume, for simplicity, that the multiplication on R satisfies $|xy| \geq |x|\,|y|$. We may then define the exponential series:

$$e^x = 1 + \frac{x}{1!} + \frac{x^2}{2!} + \cdots .$$

Then the exponential defines an isomorphism of an open additive subgroup M onto itself where M is defined by:

$$M = \{\, x \in R : v(x) < \frac{v(p)}{p-1} \,\} .$$

We may define a group law on M by setting $G(x, y)$ equal to the unique $z \in M$ such that: $e^x e^y = e^z$. It is clear from the construction of the Campbell-Hausdorff formula (L.A., Chap. 4, §7, Thm. 7,4) that this formula agrees with the Campbell-Hausdorff formula where the latter converges. Lazard has shown that the Campbell-Hausdorff formula converges in fact on M. In particular, by restriction, we obtain convergence of the Campbell-Hausdorff formula on $M \cap \mathfrak{g}$.

3) Since we have shown in Chap. 4, §8, that, when k is ultrametric, every analytic group chunk corresponds to an analytic group, we have:

Corollary. *Suppose k is ultrametric of characteristic zero. Then $\mathfrak{g}$ is the Lie algebra of an analytic group over k.*

5. Point distributions

In this section, we shall introduce "distributions whose support is concentrated at a single point". We shall use this concept as a technical tool in the next section where we prove the equivalence of the category of formal groups with the category of Lie algebras.

In considering "formal" questions, we shall assume that k is simply a commutative ring with unit or perhaps a $\mathbf{Q}$-algebra, while in considering "convergence" questions we shall assume as usual that k is a field complete with respect to a non-trivial absolute value.

1) Let X be a manifold and let $P \in X$. Recall that in Chap. 3, §7, we defined the local ring $\underline{H}_p$ of X at P and we let $\mathfrak{m}_P$ denote its maximal ideal. We showed there that if $n = \dim_P X$ then $\underline{H}_P$ is isomorphic to the ring of convergent power series in n variables. We shall give $\underline{H}_P$ the topology defined by letting the powers of $\mathfrak{m}_P$ be a basis of the neighborhoods of 0. Note that $k \subset \underline{H}_P$ inherits the discrete topology. We now define a *point distribution* on $\underline{H}_P$ to be linear form $u : \underline{H}_P \to k$ such that the following equivalent conditions are satisfied:

1. u vanishes on some power of $\mathfrak{m}_P$.
2. u is continuous on $\underline{H}_P$ with respect to the discrete topology on k.

By extension, we may also consider u as a linear form on the completion $\hat{\underline{H}}_P$ of $\underline{H}_P$. Note that $\hat{\underline{H}}_P$ is isomorphic to the formal power series ring in n variables over k.

2) Now let k be a commutative ring with unit and let $H = k[[X_1, \ldots, X_n]]$ be the formal power series ring in n variables over k. Let m be the ideal generated by $X_1, \ldots, X_n$ and, for any positive integer r, let $H_r = H/m^{r+1}$. Also let $U_r = H_r^* = L(H_r, k)$. Then, as above, we say that a linear form u on H is a *point distribution* if the following equivalent conditions are satisfied:

1. For some integer r, u vanishes on m^{r+1}, that is, u factors through the projection $H \to H_r$.
2. u is continuous on H.

Let $U \subset H^*$ be the subspace of point distributions.

Consider the projection $H \to H_r$. Dualizing, we obtain an injection $U_r \to H^*$. Then U may be identified with the union of the images of $\{U_r\}$.

Let us consider in more detail the k-module structures of H and U. By definition, H is the product: $\prod_\alpha k \cdot X^\alpha$. We may define, for each α, an element $\Delta^\alpha \in U$ by:

$$\Delta^\alpha(X^\beta) = \begin{cases} 1 & \alpha = \beta \\ 0 & \alpha \neq \beta \end{cases} .$$

The elements Δ^α of U are clearly linearly independent over k. Moreover, since each element $u \in U$ vanishes on some power of m, we may write u as a finite linear combination over k of elements of the form Δ^α. Therefore:

Lemma 1. $U = \bigoplus_\alpha k \cdot \Delta^\alpha$.

Also:

Lemma 2. $U^* = H$.

Proof. Since $U \subset H^*$, we have a map $H \to H^{**} \to U^*$. Under this map, X^α is identified with the linear form on U which is 1 on Δ^α and 0 on Δ^β, $\beta \neq \alpha$. Then:

$$U^* = L(\bigoplus_\alpha k \cdot \Delta^\alpha, k) = \prod_\alpha L(k \cdot \Delta^\alpha, k) = \prod_\alpha k \cdot X^\alpha = H \, .$$

Remark. The distribution $\varepsilon = \Delta^0$ is called the Dirac distribution.

Now let us consider the structure which the multiplication map

$$\mu : H \otimes H \to H$$

induces on U. In the ring $H \otimes H$, consider the ideal

$$\mathfrak{m} = m \otimes H + H \otimes m \, .$$

We let $H \otimes H$ have the topology defined by the powers of $\mathfrak{m}$. It then follows that μ is continuous so that μ extends to the completion $H \hat{\otimes} H$ of $H \otimes H$. Let $\underline{U}$ be the space of point distributions of $H \otimes H$ or equivalently on $H \hat{\otimes} H$. Then the dual $\mu^* : H^* \to (H \hat{\otimes} H)^*$ induces by restriction a diagonal map $\delta : U \to \underline{U}$. The terminology "diagonal" is justified by:

Lemma 3. 1. *The canonical map $U \otimes U \to \underline{U}$ is an isomorphism.*
2. (*Leibniz's formula*) $\delta(\Delta^\alpha) = \sum_{\beta+\gamma=\alpha} \Delta^\beta \otimes \Delta^\gamma$.

Proof. Let $X_i \otimes 1 = Y_i$ and $1 \otimes X_i = Z_i$ for $1 \le i \le n$. Then the canonical inclusion: $H \otimes H \to k[[Y_1, \dots, Y_n, Z_1, \dots, Z_n]] = \underline{H}$ extends to an isomorphism of $H \hat{\otimes} H$ and $\underline{H}$. Under this identification, $\underline{U}$ has a canonical basis $\Delta^{\alpha,\beta}$ defined by:

$$\Delta^{\alpha,\beta}(Y^a Z^b) = \begin{cases} 1 & \text{if } \alpha = a \text{ and } \beta = b \\ 0 & \text{otherwise} \end{cases} \, .$$

The map $U \otimes U \to \underline{U}$ identifies the basis element $\Delta^\alpha \otimes \Delta^\beta$ of $U \otimes U$ with the basis element $\Delta^{\alpha,\beta}$ of $\underline{U}$. This proves statement 1.

To prove statement 2, we consider the operation $\delta(\Delta^\alpha)$ on a typical monomial $Y^\beta Z^\gamma$. Then: $\delta(\Delta^\alpha)(Y^\beta Z^\gamma) = \Delta^\alpha(\mu(Y^\beta Z^\gamma)) = \Delta^\alpha(X^{\beta+\gamma})$. We see that this is 1 if $\beta + \gamma = \alpha$ and 0 otherwise. This proves Leibniz's formula.

Finally, since $H = k \oplus m$, we may generalize the notion of tangent vector to the algebraic situation we have been studying. Let us say that $u \in U$ is a *tangent vector* if the following equivalent conditions are satisfied:

1. $u : H \to k$ is a derivation.
2. u vanishes on k and m^2.
3. u is a primitive element for δ, that is, $\delta(u) = u \otimes 1 + 1 \otimes u$.

The equivalence of 1 and 2 follows just as in Chap. 3, §8, n°1; the equivalence of 1 and 3 follows from Leibniz formula.

6. The bialgebra associated to a formal group

We shall motivate the study of the "formal" case by first considering briefly the "convergent" case.

1) Let G be an analytic group. Let H be the local ring of G at e, m the maximal ideal of H, $\hat{H}$ the completion of H with respect to the topology defined by m, and U the point distribution on H or $\hat{H}$. Then, by the formal theory of §5, n°2, we have defined a diagonal map $\delta : U \to U \otimes U$. We shall now define a multiplication $\theta : U \otimes U \to U$ using the group law on G. Indeed, let $\phi : G \times G \to G$ denote the group law. Then, we may define a map $\sigma : \underline{H}_e \to \underline{H}_{e,e}$ by: $f \mapsto f \circ \phi$. Here $\underline{H}_e = H$ and $\underline{H}_{e,e}$ is the local ring at (e,e) on $G \times G$. We get from σ a map $\hat{\sigma} : \hat{H} \to \hat{H}_{e,e} = \hat{H} \mathbin{\hat{\otimes}} \hat{H}$. Dualizing σ, we obtain the desired map $\theta : U \otimes U \to U$. We shall see in the "formal" discussion that U is an associative algebra under θ with unit equal to the Dirac distribution.

2) We use the same notation and assumptions as in §5, n°2. In addition, we let $\underline{H} = k[[Y_1, \ldots, Y_n, Z_1, \ldots, Z_n]]$ and we let $F \in \underline{H}^n$ be a formal group law in n variables over k. Recall that we noted in the proof of Lemma 3 of §5, n°2, that $\underline{H} = H \mathbin{\hat{\otimes}} H$. We shall let $\mathfrak{m} = m \mathbin{\hat{\otimes}} H + H \mathbin{\hat{\otimes}} m$.

Since F is without constant term, given $f \in H$, we may form the composite $f \circ F \in \underline{H}$ (Bourbaki, *Alg.*, Chap. 4, §5, n°5). Let $\sigma : H \to \underline{H}$ denote the map: $f \mapsto f \circ F$. Then, $\sigma(m^r) \subset \mathfrak{m}^r$ so that σ is continuous with respect to the topologies defined by m and $\mathfrak{m}$. Hence σ induces a map $\theta : U \otimes U \to U$ by dualizing.

Lemma 1. *θ makes U into an associative algebra with unit equal to the Dirac distribution.*

Proof. We shall be concerned with the power series ring $\underline{H} = k[[Y, Z]]$ and with the power series ring $\mathbf{H} = k[[Y, Z, W]]$ where $W = (W_1, \ldots, W_n)$. We shall let Δ_Y^α, Δ_Z^β, Δ_W^γ denote the elements dual to the monomials Y^α, Z^β, W^γ respectively. Then, for example, the product $\Delta_Y^\alpha \Delta_Z^\beta \Delta_W^\gamma$ is dual to $Y^\alpha Z^\beta W^\gamma$.

We now prove the associative law for θ. Let $\Delta^\alpha \otimes \Delta^\beta \otimes \Delta^\gamma$ be a typical basis element of $U \otimes U \otimes U$. Let $f \in H$. Then:

$$\theta(\Delta^\alpha, \theta(\Delta^\beta, \Delta^\gamma))f = \Delta_Y^\alpha \Delta_Z^\beta \Delta_W^\gamma (f \circ F(Y, F(Z, W)))$$
$$\theta(\theta(\Delta^\alpha, \Delta^\beta), \Delta^\gamma)f = \Delta_Y^\alpha \Delta_Z^\beta \Delta_W^\gamma (f \circ F(F(Y, Z), W)) .$$

The associativity of θ therefore follows from the associativity of F.

We conclude the lemma by showing that the Dirac distribution ε is a unit for the multiplication in U. Let $\Delta^\alpha \in U$ and $f \in H$. Then:

$$\theta(\varepsilon, \Delta^\alpha)f = \varepsilon_Y \Delta_Z^\alpha (f \circ F(Y, Z)) = \Delta_Z^\alpha (f \circ F(0, Z)) = \Delta_Z^\alpha (f(Z)) .$$

This shows that $\theta(\varepsilon, \Delta^\alpha) = \Delta^\alpha$. Hence, ε is a left unit, and it is shown in a similar way that ε is a right unit.

Lemma 2. *The diagonal map $\delta : U \to U \otimes U$ is an algebra homomorphism.*

Proof. We must show that the following diagram is commutative:

$$\begin{array}{ccccc} (U \otimes U) \otimes (U \otimes U) & & \xrightarrow{\alpha} & & (U \otimes U) \otimes (U \otimes U) \\ \uparrow{\scriptstyle \delta \otimes \delta} & & & & \downarrow{\scriptstyle \theta \otimes \theta} \\ U \otimes U & \xrightarrow{\theta} & U & \xrightarrow{\delta} & U \otimes U \end{array} ,$$

where $\alpha(x \otimes y \otimes z \otimes t) = x \otimes z \otimes y \otimes t$.

Since all the modules appearing in the diagram are free, it suffices to show that the dualized diagram is commutative:

$$\begin{array}{ccccc} (H \mathbin{\hat\otimes} H) \mathbin{\hat\otimes} (H \mathbin{\hat\otimes} H) & & \xleftarrow{\alpha'} & & (H \mathbin{\hat\otimes} H) \mathbin{\hat\otimes} (H \mathbin{\hat\otimes} H) \\ {\scriptstyle \mu \hat\otimes \mu}\downarrow & & & & \uparrow{\scriptstyle \sigma \hat\otimes \sigma} \\ H \mathbin{\hat\otimes} H & \xleftarrow{\hat\sigma} & \hat H & \xleftarrow{\mu} & H \mathbin{\hat\otimes} H \end{array} .$$

Since all maps are continuous with respect to the appropriate topologies, it suffices to check commutativity on elements of $H \mathbin{\hat\otimes} H$ of the form $f \otimes g$ where $f, g \in H$. But:

$$\begin{aligned} \hat\sigma(\mu(f \otimes g)) &= \hat\sigma(f \cdot g) = (f \cdot g) \circ F = (f \circ F) \cdot (g \circ F) \\ &= (\mu \mathbin{\hat\otimes} \mu) \circ \alpha'(f \circ F \otimes g \circ F) = (\mu \mathbin{\hat\otimes} \mu) \circ \alpha' \circ (\sigma \mathbin{\hat\otimes} \sigma)(f \otimes g)) . \end{aligned}$$

This proves the lemma.

The space U together with the diagonal map $\delta : U \to U \otimes U$ and the multiplication map $\theta : U \otimes U \to U$ is called the *bialgebra* associated to the formal group F. We use the notations: $\theta(u, v) = u * v$ and $[u, v]_* = u * v - v * u$.

Lemma 3. *For all α and β, $\Delta^\alpha * \Delta^\beta = \binom{\alpha+\beta}{\alpha} \Delta^{\alpha+\beta} + E_{\alpha,\beta}$, where the error term $E_{\alpha,\beta}$ is a linear combination of Δ^γ with $0 < |\gamma| < |\alpha + \beta|$.*

Proof. We must show that if $|\gamma| \geq |\alpha + \beta|$ then $\Delta^\alpha * \Delta^\beta$ and $\binom{\alpha+\beta}{\alpha} \Delta^{\alpha+\beta}$ agree on X^γ and that $E_{\alpha,\beta}$ vanishes on k. But:

$$X^\gamma \circ F(Y,Z) = (F(Y,Z))^\gamma = (Y+Z)^\gamma + O(d^0 > |\gamma|) ,$$

and

$$(Y+Z)^\gamma = \sum_{\lambda+\mu=\gamma} \binom{\gamma}{\lambda} Y^\lambda Z^\mu .$$

Since $\Delta^\alpha * \Delta^\beta(X^\gamma) = \Delta^\alpha_Y \Delta^\beta_Z(X^\gamma \circ F(Y,Z))$, one finds:

$$\begin{aligned} \Delta^\alpha * \Delta^\beta(X^\gamma) &= \begin{cases} \binom{\alpha+\beta}{\alpha} & \text{if } \gamma = \alpha+\beta \\ 0 & \text{otherwise} \end{cases} \\ &= \binom{\alpha+\beta}{\alpha} \Delta^{\alpha+\beta}(X^\gamma) . \end{aligned}$$

This proves the assertion about the action of $\Delta^\alpha * \Delta^\beta$ and $\binom{\alpha+\beta}{\alpha}\Delta^{\alpha+\beta}$ in X^γ if $|\gamma| > |\alpha+\beta|$. We also see that both elements give 1 on X^0 when $\alpha+\beta=0$ and 0 otherwise so that $E_{\alpha,\beta}$ always vanishes on k.

Let $\mathfrak{g}$ be the Lie algebra associated to the formal group F (§1). Then, the vector space underlying $\mathfrak{g}$ is k^n. We let D_i be the i-th standard basis vector of k^n and define $\psi : \mathfrak{g} \to U$ by: $\psi(D_i) = \Delta^{\delta_i}$. Then, by definition, ψ is a linear isomorphism of $\mathfrak{g}$ onto the set of $u \in U$ such that u vanishes on k and m^2.

Lemma 4. *Let $x, y \in \mathfrak{g}$ and let $f \in H$ be a linear function. Then*

$$\psi([x,y])f = f(B(x,y) - B(y,x)) .$$

Proof. Since, by definition, $[x,y] = B(x,y) - B(y,x)$, we must prove:

$$\psi([x,y])f = f([x,y]) .$$

Then, since both sides of this equation are trilinear in x, y, f, it suffices to consider the case when $x = D_i$, $y = D_j$, and $f = X_k$. Then, both sides of the equation reduce to the structure constant $\gamma^k_{ij}(D_1, \dots, D_n)$, which proves the lemma.

Theorem 1. *ψ is a Lie algebra homomorphism: $\psi([x,y]) = [\psi x, \psi y]_*$.*

Proof. We know that $\psi([x,y])$ vanishes on k and m^2 and it follows from Lemma 3 that $[\psi x, \psi y]_*$ vanishes on k and m^2. Hence, to prove the desired equality, it suffices to show that $\psi([x,y])$ and $[\psi x, \psi y]_*$ agree on a set of coset representatives of m/m^2, for example, the linear functions. Let $f \in H$ be a linear function. Then:

$$\begin{aligned} [\psi x, \psi y]_* f &= (\psi x \otimes \psi y - \psi y \otimes \psi x)(f \circ F(X,Y)) \\ &= (\psi x \otimes \psi y - \psi y \otimes \psi x)(f \cdot X + f \cdot Y + f \cdot B(X,Y) + \cdots) \\ &= (\psi x \otimes \psi y - \psi y \otimes \psi x)(f \cdot B(X,Y)) \\ &= f(B(x,y) - B(y,x)) . \end{aligned}$$

The desired equality follows now from Lemma 4.

Henceforth, we assume that k is a $\mathbf{Q}$-algebra. Par abus de notation, we let $\psi : U\mathfrak{g} \to U$ denote the map induced by $\psi : \mathfrak{g} \to U$ (L.A., Chap. 3, §1, Def. 1.1).

Theorem 2. *$\psi : U\mathfrak{g} \to U$ is a bialgebra isomorphism.*

We have defined filtrations on $U\mathfrak{g}$ (L.A., Chap. 3, §4) and on U (§5, n°2). We verify immediately that ψ is compatible with these filtrations and that the filtration on $U\mathfrak{g}$ (resp. U) is separated, exhaustive, and makes $U\mathfrak{g}$ (resp. U) discrete. Hence, by Bourbaki, *Alg. Comm.*, Chap. 3, §2, n°8, Cor. 3 of Thm. 1, it suffices to show that $\mathrm{gr}(\psi)$ is bijective to prove that ψ is bijective.

We know from the Poincaré-Birkhoff-Witt Theorem (L.A., Chap. 3, §4, Thm. 4.3) that $U\mathfrak{g}$ is a free k-module with basis $D^\alpha = D_1^{\alpha_1} \cdots D_n^{\alpha_n}$. Since k is an algebra over $\mathbf{Q}$, we know that U is a free k-module with basis $\underline{D}^\alpha = \frac{1}{\alpha!}\Delta^\alpha$. By Lemma 2, $\psi(D^\alpha)$ and $\underline{D}^\alpha$ agree in $\mathrm{gr}(U)$. Hence $\mathrm{gr}(\psi)$ is bijective so that ψ is bijective.

We know that ψ is an algebra homomorphism so that to show it is a bialgebra homomorphism it remains to show that the following diagram commutes:

$$\begin{array}{ccc} U\mathfrak{g} & \xrightarrow{\Delta} & U\mathfrak{g} \otimes U\mathfrak{g} \\ \psi\downarrow & & \downarrow\psi\otimes\psi \\ U & \xrightarrow{\delta} & U \otimes U \end{array} .$$

We know, by L.A., Chap. 3, §5, Prop. 5.2, by §5, n°2, Lemma 3, and by Lemma 2 above that:

1. Δ and δ are algebra homomorphisms.
2. For $u \in \mathfrak{g} \subset U\mathfrak{g}$, $(\psi \otimes \psi)\Delta(u) = \psi u \otimes 1 + 1 \otimes \psi u = \delta \circ \psi(u)$.

Since, $\mathfrak{g}$ generates the algebra $U\mathfrak{g}$ (L.A., Chap. 3, §4, Prop. 4.1), $\psi \circ \Delta = \delta \circ \psi$ on all of $U\mathfrak{g}$, as desired.

Theorem 3. *Let T : (FG) $\to$ (LA) be the functor from the category of formal groups over k to the category of finite dimensional Lie algebras over k defined in §1. Then, T is an equivalence of categories, that is:*

1. *For $F_1, F_2 \in$ (FG), the map:* $\mathrm{Hom}(F_1, F_2) \to \mathrm{Hom}(TF_1, TF_2)$ *is a bijection.*

2. *Given $\mathfrak{g} \in$ (LA), there exists $F \in$ (FG) such that TF is isomorphic to $\mathfrak{g}$. (Recall that k is supposed to be a $\mathbf{Q}$-algebra.)*

Proof. 1. For $i = 1, 2$, let F_i be the formal group law in n_i variables over k, H_i be the formal power series ring: $k[[X_1, \ldots, X_{n_i}]]$, U_i be the bialgebra of point distribution on H_i, $\mathfrak{g}_i$ be the Lie algebra associated to F_i, and $\underline{U}_i$ be the universal algebra of $\mathfrak{g}_i$; use the notations:

a. $\mu_i : H_i \hat{\otimes} H_i \to H_i$
$\theta_i : U_i \otimes U_i \to U_i$ multiplications maps
$\underline{\theta}_i : \underline{U}_i \otimes \underline{U}_i \to \underline{U}_i$

b. $\sigma_i : H_i \to H_i \hat{\otimes} H_i$
$\delta_i : U_i \to U_i \otimes U_i$ diagonal maps
$\underline{\delta}_i : \underline{U}_i \to \underline{U}_i \otimes \underline{U}_i$

c. $\psi_i : \underline{U}_i \to U_i$: the isomorphism of Theorem 2.

Then, given a Lie algebra homomorphism $t : \mathfrak{g}_1 \to \mathfrak{g}_2$, we want to show there exists a unique formal group homomorphism $\tau : F_1 \to F_2$, such that $T(\tau) = t$.

We will show that the map: $\mathrm{Hom}(F_1, F_2) \to \mathrm{Hom}(TF_1, TF_2)$ can be decomposed into a series of maps each of which is a bijection.

0) We begin with a preliminary step. Let:

$\mathrm{Hom}_{\mathrm{AL}}(H_2, H_1)$ = continuous algebra homomorphisms mapping m_2 into m_1 (such a homomorphism will be called *admissible* – it is continuous for the natural topologies of H_2 and H_1).

S = space of $\tau = (\tau_1, \ldots, \tau_{n_2}) \in H_1^{n_2}$ such that $\tau_i(0) = 0$ for all i.

Let $\tau \in S$. Then, given $g \in H_2$, we may form the composite $g \circ \tau$, and the map $\phi_\tau : H_2 \to H_1$ defined by $g \mapsto g\tau$ is an admissible algebra homomorphism (Bourbaki, *Alg.*, Chap. 4, §5, n°5, Prop. 3).

Lemma 5. *The map $\tau \mapsto \phi_\tau$ is a bijection of S onto* $\mathrm{Hom}_{\mathrm{AL}}(H_2, H_1)$.

Proof. This map is injective since $\phi_\tau(X_i) = \tau_i$. To prove that it is surjective, let $\phi : H_2 \to H_1$ be an admissible algebra homomorphism and let τ be defined by $\tau_i = \phi(X_i)$. Then, since ϕ and ϕ_τ are algebra homomorphisms, ϕ and ϕ_τ agree on $k[X_1, \ldots, X_{n_2}]$. Since this subring is dense in H_2, $\phi = \phi_\tau$ on H_2 by continuity.

1) Let:

$$\mathrm{Hom}_{\mathrm{BA}}(H_2, H_1) = \text{admissible bialgebra homomorphisms} .$$

Lemma 6. *Let $\tau \in S$. Then:*

$$\tau \in \mathrm{Hom}_{\mathrm{FG}}(F_1, F_2) \iff \phi_\tau \in \mathrm{Hom}_{\mathrm{BA}}(H_2, H_1) .$$

Proof. The statement that $\phi_\tau \in \mathrm{Hom}_{\mathrm{BA}}(H_2, H_1)$ amounts by Lemma 5 to the commutativity of the following diagram:

$$\begin{array}{ccc} H_2 & \xrightarrow{\delta_2} & H_2 \otimes H_2 \\ \phi\downarrow & & \downarrow\phi_\tau\otimes\phi_\tau \\ H_1 & \xrightarrow{\delta_1} & h_1 \otimes H_1 \end{array} .$$

The diagram commutes if and only if for all $g \in H_2$:

$$g \circ \tau \circ F_1(X,Y) = \delta_1 \circ \phi_\tau(g) = (\phi_\tau \otimes \phi_\tau) \circ \delta_2(g) = g \circ F_2(\tau X, \tau Y)\ .$$

This equation holds for all $g \in H_2$ if and only if $\tau \circ F_1(X,Y) = F_2(\tau X, \tau Y)$, that is, if and only if τ is a formal group homomorphism.

2) Let:

$$\mathrm{Hom}_{\mathrm{BA}}(U_1, U_2) = \text{bialgebra homomorphisms}$$
$$\mathrm{Hom}_{\mathrm{BA}}(\underline{U}_1, \underline{U}_2) = \text{bialgebra homomorphisms}\ .$$

Lemma 7. *Dualization defined an isomorphism*

$$\mathrm{Hom}_{\mathrm{BA}}(H_2, H_1) \approx \mathrm{Hom}_{\mathrm{BA}}(U_1, U_2)\ .$$

Proof. Given $\phi : H_2 \to H_1$, ϕ an admissible bialgebra homomorphism, and $u \in U_1$, we obtain $u \circ \phi \in U_2$. This follows since the U_i are the continuous duals of the H_i. Hence we have a map: $\mathrm{Hom}_{\mathrm{BA}}(H_2, H_1) \to \mathrm{Hom}_{\mathrm{BA}}(U_1, U_2)$.

Given $\psi : U_1 \to U_2$ and $g \in H_2$, we obtain $g \circ \psi \in H_1$. This follows from Lemma 2 of §5, n°2. The fact that ψ is a bialgebra homomorphism shows (by duality) that $g \mapsto g \circ \psi$ is a bialgebra homomorphism. Moreover, since $\psi(\varepsilon_1) = \varepsilon_2$, $g \in m_2 \Rightarrow g \circ \psi \in m_1$. Hence $g \mapsto g \circ \psi$ is admissible, and we have defined a map: $\mathrm{Hom}_{\mathrm{BA}}(U_1, U_2) \to \mathrm{Hom}_{BA}(H_1, H_1)$.

It is easily checked that the two maps we have defined are inverse to one another.

3) By Theorem 2:

Lemma 8. *The maps ψ_i define an isomorphism:*

$$\mathrm{Hom}_{\mathrm{BA}}(U_1, U_2) \to \mathrm{Hom}_{\mathrm{BA}}(\underline{U}_1, \underline{U}_2)\ .$$

4) From the definition of the universal algebra and the definition of the diagonal map for the universal algebra, we have:

Lemma 9. $\mathrm{Hom}_{\mathrm{BA}}(\underline{U}_1, \underline{U}_2) = \mathrm{Hom}_{\mathrm{LA}}(\mathfrak{g}_1, \mathfrak{g}_2)$.

5) It remains to put the bijections of Lemmas 6, 7, 8, 9 together and to see that they give the functor T. Let $\tau \in \mathrm{Hom}_{\mathrm{FG}}(F_1, F_2)$ and write:

$$\tau(X) = \sum_i t_i(X)\ ,$$

where $t_i(X)$ is homogeneous of degree i. Then $T(\tau) = t_1$. Let $t \in \mathrm{Hom}_{\mathrm{LA}}(\mathfrak{g}_1, \mathfrak{g}_2)$ be the element corresponding to τ under the above bijections. We must show that $t = t_1$.

Let $u \in \mathfrak{g}_1$. To see whether $t(u) = t_1(u)$, it suffices to test by applying linear functions $g \in H_2$ to both sides. We have:

$$g(t(u)) = t(u)g = u(\phi_\tau(g)) = g(g \circ \tau) = u\Big(\sum_i gt_i\Big) = u(gt_1) \; ;$$

$$g(t_1(u)) = u(gt_1) \; .$$

The first line is simply unwinding the isomorphisms of Lemmas 6, 7, 8, 9, while the second is the identification we have made of $\mathfrak{g}_i$ with a subset of U_i via ψ_i. Comparing, we obtain the desired equality.

2. We associate to $\mathfrak{g} \in (\mathrm{LA})$ the Campbell-Hausdorff formal group with $\mathfrak{g}$ as Lie algebra. The details of this association were essentially carried out in the first part of the proof of Theorem 1 of §4. The proof there was entirely "formal" and used only the fact that k is a $\mathbf{Q}$-algebra.

Remark. One could also prove part 2) of Theorem 3 by showing directly that the dual of the bialgebra $U\mathfrak{g}$ is isomorphic (as a k-algebra) to a formal power series ring $k[[X_1, \ldots, X_n]]$, with $n = \dim \mathfrak{g}$; the diagonal structure of this algebra is given by an $F(X, Y)$ which is the desired formal group law.

7. The convergence of formal homomorphisms

We assume that k is a complete field with respect to a non-trivial absolute value, and that $\operatorname{char} k = 0$.

Theorem 1. *For $i = 1, 2$, let G_i be an analytic group chunk and F_i be the formal group law induced in k^{n_i} by a chart c_i of G_i at e_i. Let $\tau : F_1 \to F_2$ be a formal group homomorphism. Then, τ is convergent, that is, τ induces a local homomorphism of group chunks $\hat{\tau} : G_1 \dashrightarrow G_2$.*

Proof. For $i = 1, 2$, F_i is convergent since it is obtained by passage to local coordinates from the convergent multiplication law on G_i; since the conclusion of the theorem is local, we may assume that G_i is an open neighborhood of 0 in k^{n_i} with multiplication defined by F_i.

1) Special Case: $G_1 = k$, $F_1 =$ "+", and $G_2 = G$, $F_2 = F$.

To say that τ is a formal group homomorphism reduces, in this case, to the formal equations:

$$\tau(s+t) = F(\tau(s), \tau(t)) \quad \text{and} \quad \tau(0) = 0 \; .$$

Differentiating formally with respect to t and setting $t = 0$, we find that τ satisfies the following formal differential equation:

$$\tau'(s) = D_2F(\tau(s), 0)\tau'(0) \; .$$

Since $D_2F(\tau(0), 0) = D_2F(0, 0) = \mathrm{Id}_{k^n}$, the above equation is formally consistent at $s = 0$. Let $\phi(X) = D_2F(X, 0)\tau'(0)$ where $\tau'(0)$ is any fixed vector in k^n. Then, the convergence of τ is a consequence of the following theorem which we prove in the appendix to this chapter.

Theorem 2. *Suppose $\phi = (\phi_1, \ldots, \phi_n)$ is a system of n convergent power series in n variables. Then, the formal differential equation: $\tau'(s) = \phi(\tau(s))$, $\tau(0) = 0$, possesses a unique formal solution. This solution is convergent.*

2) General Case: Let F be a convergent formal group law corresponding to a group chunk G and let $X \in \mathfrak{g} = L(G)$. Then there is a unique Lie algebra homomorphism $L_X : k \to \mathfrak{g}$ such that $L_X(1) = X$. By Theorem 3 of §6, there is a unique formal homomorphism $\phi_X : k^+ \to F$ such that $\phi'_X(0) = X$. By case 1), ϕ_X is convergent.

Let $\mathfrak{g}_i = L(G_i)$ and let $t = L(\tau) : \mathfrak{g}_1 \to \mathfrak{g}_2$. Then the construction of ϕ_X has the following functorial property:

For $X \in \mathfrak{g}_1$ and $Y = t(X)$: $\phi_Y = \tau \circ \phi_X$ formally.

For convenience, set $n = n_1$ and $m = n_2$. Then, choose a basis $\{X_\mu\}$ of $\mathfrak{g}_1$ and let $Y_\mu = t(X_\mu)$. Define local morphisms:

$$k^n \begin{array}{l} \overset{\phi_1}{-\!-\!\!\rightarrow} G_1 \\ \underset{\phi_2}{-\!-\!\!\rightarrow} G_2 \end{array},$$

by the formulae:

$$\begin{aligned} \phi_1(t_1, \ldots, t_n) &= \phi_{X_1}(t_1) \cdots \phi_{X_n}(t_n) \\ \phi_2(t_1, \ldots, t_n) &= \phi_{Y_1}(t_1) \cdots \phi_{Y_n}(t_n) \,. \end{aligned}$$

The map $L(\phi_1) : k^n \to \mathfrak{g}_1$ is just the isomorphism of k_n with $\mathfrak{g}_1$ defined by sending the μ-th standard basis vector of k^n onto X_μ. Hence ϕ_1 is étale at 0 and hence is a local isomorphism in a neighborhood of 0.

Now, formally, $\tau \circ \phi_1 = \phi_2$, hence, formally, $\tau = \phi_2 \circ \phi_1^{-1}$. But the right hand side of this equation is convergent by what we have just shown. Hence τ is convergent, as desired.

Corollary 1. *Let G_1 and G_2 be analytic group chunks such that $L(G_1)$ and $L(G_2)$ are isomorphic. Then:*

1. *Any isomorphism of $L(G_1)$ with $L(G_2)$ induces a local isomorphism of G_1 with G_2.*
2. *In the ultrametric case, G_1 and G_2 have open subgroups which are isomorphic.*

Proof. 1. This follows from Theorem 1 and from Theorem 3, §6.

2. This follows from 1) and from the theorem of Chap. 4, §8.

Corollary 2. *Let G be an analytic group chunk, $\mathfrak{g} = L(G)$, and* CH($\mathfrak{g}$) = *Campbell-Hausdorff group chunk associated to $\mathfrak{g}$ (§4). Then, there exists a unique local isomorphism* exp : CH($\mathfrak{g}$) $\to G$ *such that* $L(\exp) = \mathrm{id}$.

Remark. Suppose $G = \mathrm{GL}(V)$ where V is a vector space over k. Then let:

$$e^x = 1 + \frac{x}{1!} + \frac{x^2}{2!} + \cdots .$$

We content that $\exp(x) = e^x$ in a neighborhood of 0 in $\mathfrak{g} = E(V)$. Indeed, by the construction of the Campbell-Hausdorff formula, e^X defines a formal homomorphism from $\mathrm{CH}(\mathfrak{g})$ to G and we have $L(e^X) = \mathrm{Id}$. Hence, we obtain: $\exp(X) = e^X$, by uniqueness.

Now let us study the map exp in the case $k = \mathbf{R}$ or $\mathbf{C}$ and G is an analytic group. Since $f_n(X) = nX$ in $\mathrm{CH}(\mathfrak{g})$, we will be able to extend exp to all of $\mathfrak{g}$. Indeed, let $x \in \mathfrak{g}$. Then, since $k = \mathbf{R}$ or $\mathbf{C}$, for some integer $n > 0$, $\frac{1}{n}x \in \mathrm{Dom}(\exp)$. Define:

$$(*) \qquad \exp(x) = \exp(\frac{1}{n}x)^n .$$

We obtain the same definition if we replace n by a multiple mn of n. This shows that we obtain a unique definition independent of the choice of n, since any two choices may be compared by means of their product.

For fixed x_0 and some n, the formula $(*)$ is valid in a neighborhood of x_0, which shows that exp is analytic. We know that exp is étale at 0 but in general exp is not everywhere étale and is not bijective.

Corollary 3. *Suppose $k = \mathbf{R}$ or $\mathbf{C}$. Let G be an analytic group over k and let $\mathfrak{g} = L(G)$. Then:*

1. *Suppose G is connected and $\mathfrak{g}$ is abelian. Then, G is abelian.*
2. *More generally, if $\mathfrak{g}$ is abelian, then G has an open abelian subgroup.*

Proof. Clearly, $1 \Rightarrow 2$. To show 1, we note first that, since $\mathfrak{g}$ is abelian, $\mathrm{CH}(\mathfrak{g}) = \mathfrak{g}$ with the additive group structure. We content exp defines a group homomorphism. Indeed, given $x, y \in \mathfrak{g}$, choose an integer n so large that $\frac{1}{n}x$, $\frac{1}{n}y$, $\frac{1}{n}(x+y)$ lie in the domain where exp is a local homomorphism. Then:

$$\exp(\frac{1}{n}x)\exp(\frac{1}{n}y) = \exp(\frac{1}{n}(x+y)) = \exp(\frac{1}{n}y)\exp(\frac{1}{n}x) .$$

Noting the commutability of $\exp(\frac{1}{n}x)$ and $\exp(\frac{1}{n}y)$, we obtain, by raising to n-th powers:

$$\exp(x)\exp(y) = \exp(x+y) .$$

It follows that $\exp(\mathfrak{g})$ is an open abelian subgroup of G, hence is equal to G, since G is connected.

Corollary 4. *Suppose k is ultrametric and let A be the corresponding valuation ring. Let G be an analytic group over k of dimension n such that $\mathfrak{g} = L(G)$ is abelian. Then, G has an open abelian subgroup isomorphic to A^n.*

Proof. We have that $\mathrm{CH}(\mathfrak{g})$ is the additive group $\mathfrak{g}$ and that $\mathrm{CH}(\mathfrak{g})$ contains an open subgroup isomorphic to an open subgroup of G, whence the result.

Example of Corollary 4. $G =$ groups of points of an abelian variety defined over $\mathbf{Q}_p$.

8. The third theorem of Lie

Throughout this section, $k = \mathbf{R}$ or $\mathbf{C}$.

Theorem 1. *Let G be a connected, simply connected analytic group over k. Let G' be any analytic group over k. Let $\mathfrak{g} = L(G)$ and $\mathfrak{g}' = L(G')$. Then, the map: $\mathrm{Hom}_{\mathrm{AG}}(G, G') \to \mathrm{Hom}_{\mathrm{LA}}(\mathfrak{g}, \mathfrak{g}')$ is bijective.*

Proof. Let $t : \mathfrak{g} \to \mathfrak{g}'$ be a Lie algebra homomorphism. We must show that $t = L(\phi)$ for a unique analytic group homomorphism $\phi : G \to G'$. We know that t extends uniquely to a local homomorphism $f : G \dashrightarrow G'$ (§6, Thm. 3; §7, Thm. 1). Then, the graph $\Gamma_f \subset G \times G'$ of f is an analytic subgroup chunk. Let (H, i) be the analytic group generated by Γ_f (Chap. 4, §4). Consider the diagram:

$$\begin{array}{ccc} H & \xrightarrow{i} & G \times G' \\ & {\scriptstyle\psi}\searrow & \downarrow {\scriptstyle \mathrm{pr}_1} \\ & & G \end{array}$$

Then, ψ is a local isomorphism, and, since H is connected and G is simply connected, ψ is an isomorphism onto an open subgroup of G. Since G is connected, ψ is surjective. Let $\phi = \mathrm{pr}_2 \circ i \circ \psi^{-1}$. Then, ϕ agrees with f in a neighborhood of e. This shows existence. Uniqueness follows since two homomorphisms ϕ_1 and ϕ_2 whose derivative at e is equal to t must locally agree with f so that the set of points on which they agree is open and closed, therefore, equal to all of G.

Theorem 2. *The category of connected, simply connected analytic groups over $k = \mathbf{R}$ or $\mathbf{C}$ is equivalent to the category of finite dimensional Lie algebras.*

Proof. We have proved in Theorem 1 the required bijection on maps. What remains to be done is to show:

Theorem 3 (Third Theorem of Lie). *For any finite dimensional Lie algebra $\mathfrak{g}$, there exists a connected and simply connected analytic group G such that $L(G) = \mathfrak{g}$.*

Proof. This theorem was first proved by Elie Cartan. We shall sketch Cartan's proof after giving a shorter proof based on the powerful Theorem of Ado.

We remark at the outset that it suffices to find an analytic group G such that $L(G) = \mathfrak{g}$ since taking the connected component H of e in G and then taking the simply covering group of H, we obtain the desired connected, simply connected analytic group with Lie algebra $\mathfrak{g}$.

Proof 1. We quote Ado's Theorem (Bourbaki, *Alg. de Lie*, Chap. 1, §7, n°3, Thm. 3, or Jacobson, *Lie Algebras*, Chap. 6, §2, p. 202):

Theorem (Ado). *Every finite dimensional Lie algebra has a faithful finite dimensional representation.*

Now, let H be the Campbell-Hausdorff group chunk corresponding to $\mathfrak{g}$ and let $t : \mathfrak{g} \to E(V)$ be a faithful representation. Then t induces a local homomorphism $f : H \dashrightarrow \mathrm{GL}(V)$. Since t is faithful, f is an immersion at e, that is, H corresponds to a subgroup chunk of $\mathrm{GL}(V)$. But then, H is equivalent to an analytic group (Chap. 4, §4).

Proof 2. 1) The theorem is true in the following cases:

1. $\mathfrak{g}$ is semisimple.
2. $\mathfrak{g}$ is abelian.

In case 1, $\mathrm{ad} : \mathfrak{g} \to E(\mathfrak{g})$ is injective. We may apply the method of proof 1 without having to invoke Ado's Theorem. Alternatively, we note that $\mathrm{Im}(\mathrm{ad}) = \mathrm{Der}(\mathfrak{g})$. Hence, $\mathfrak{g} = L(\mathrm{Aut}(\mathfrak{g}))$ (§3, n°4, 4).

Case 2 is trivial since we may take $G =$ additive group of $\mathfrak{g}$.

2) General case:

We use induction on $\dim \mathfrak{g}$. If $\mathfrak{g}$ falls in cases 1 or 2, we are done. Otherwise, we know that $\mathfrak{g}$ is a semi-direct product $\mathfrak{g}_1 \tilde{\times} \mathfrak{g}_2$, where $\mathfrak{g}_1$ is an ideal of $\mathfrak{g}$, $\mathfrak{g}_2$ a subalgebra of $\mathfrak{g}$, and $\dim \mathfrak{g}_i < \dim \mathfrak{g}$ (cf. L.A., Chap. 6, §4).

Let $\phi : \mathfrak{g}_2 \to \mathrm{Der}(\mathfrak{g}_1)$ define the semi-direct product structure on $\mathfrak{g}_1 \tilde{\times} \mathfrak{g}_2$. By induction, let G_i be a connected, simply connected analytic group such that $\mathfrak{g}_i = L(G_i)$, $i = 1, 2$. We will show that $\mathfrak{g} = L(G)$ where G is a semi-direct product of G_1 and G_2.

The main steps in the proof are:

1. $\mathrm{Der}(\mathfrak{g}_1) = L(A_1)$ where $A_1 = \mathrm{Aut}(\mathfrak{g}_1)$ (§3, n°4, 4).
2. $A_1 = \mathrm{Aut}(G_1)$, by Theorem 1.
3. A_1 acts analytically on G_1.

Indeed, given $a \in A_1$ and $g \in G$, we want to find a neighborhood N of a in A_1 and U of g in G_1 such that the action: $N \times U \to G$ is analytic. Let $W_1 \subset \mathfrak{g}_1$ be a neighborhood of 0 in $\mathfrak{g}$ on which the Campbell-Hausdorff formula converges. Choose neighborhoods N of a in A_1 and W_2 of 0 in W_1 so that the action of $N \subset \mathrm{Aut}(\mathfrak{g}_1)$ on W_2 takes values on W_1. Let $V = \exp(W_2)$. We then have that the action of $N \subset \mathrm{Aut}(G_1)$ on V is induced by the action of $N \subset \mathrm{Aut}(\mathfrak{g}_1)$ on W_2. Since the elements of $\mathrm{Aut}(\mathfrak{g}_1)$ act linearly on the

Campbell-Hausdorff formula, the action of N on W_2 is analytic. Thus, the action of N on V is analytic.

Now, since G_1 is connected, there exists an integer $n > 0$ such that $g \in V^n$, where V_n denotes the set of products of n elements of V. We may assume that V is open. Then V_n, which is the union of translates of V, is open. Take $U = V^n$. Let $\theta : A_1 \times G_1 \to G_1$ be the action of A_1 on G_1; write $\theta(b,h) = bh$. We must show that θ is analytic on $N \times U$. Let $V^{(n)}$ be the n-fold product of V and let $\mu : V^{(n)} \to U$ be multiplication. Also, let $\underline{\theta} : N \times V^{(n)} \to G$ be defined by:

$$\underline{\theta}(b, g_1, \ldots, g_n) = (bg_1)\cdots(bg_n) .$$

Then, the following diagram is commutative:

$$\begin{array}{ccc} N \times V^{(n)} & \xrightarrow{\underline{\theta}} & G \\ {\scriptstyle 1\times\mu}\downarrow & \nearrow{\scriptstyle \theta} & \\ N \times U & & \end{array} .$$

Since $\underline{\theta}$ is analytic, to prove that θ is analytic, it suffices to remark that $\mu : V^{(n)} \to U$ is a surjective submersion.

4. Since G_2 is simply connected and connected, ϕ induces $\psi : G_2 \to A_1$. We define a semi-direct product structure on the set $G_1 \times G_2$ by:

$$(0,h)(g,0)(0,h)^{-1} = (\psi(h)g, 0) .$$

This group structure is analytic since ψ is analytic and since A_1 acts on G_1 analytically.

It is now a simple verification that $L(G_1 \tilde{\times} G_2) = \mathfrak{g}$.

Theorem 4. *Let G be a connected and simply connected analytic group. Let $\mathfrak{g} = L(G)$ and let $\mathfrak{h} \subset \mathfrak{g}$ be an ideal. Then:*

1. *There is a closed connected analytic subgroup H of G such that $L(H) = \mathfrak{h}$.*
2. *H is simply connected.*

Proof. 1. Let K be an analytic group such that $L(K) = \mathfrak{g}/\mathfrak{h}$. The projection of $\mathfrak{g}$ on $\mathfrak{g}/\mathfrak{h}$ induces an analytic homomorphism $\phi : G \to K$ since G is connected, and simply connected. Take H to be the connected component of e in $\operatorname{Ker}\phi$. Then H has the required properties.

2. We use the fact that since G/H is an analytic group then $\pi_2(G/H) = 0$. Then, in the homotopy exact sequence, we have:

$$\begin{array}{ccccc} \cdots \longrightarrow \pi_2(G/H) & \longrightarrow & \pi_1(H) & \longrightarrow & \pi_1(G) \\ \| & & & & \| \\ 0 & & & & 0 \end{array}$$

Hence, $\pi_1(H) = 0$.

Remark. It seems likely that no "simple" proof of Lie's Third Theorem exists. For, if such a proof did exist, unless it made essential use of the local compactness of **R** and **C**, it would extend to Banach analytic groups. But in the Banach space setting, the Third Theorem is *false* (as remarked by van Est and others). Indeed, Theorem 4, 1 itself (which is a formal consequence of Theorem 3) is false:

The example is the following. Take $G = \mathrm{GL}(H) \times \mathrm{GL}(H)$ where H is an infinite dimensional Banach space. It is known that $\mathrm{GL}(H)$ is connected and simply connected. The center of G contains $\mathbf{C}^* \times \mathbf{C}^*$ and hence $S^1 \times S^1$. We let $Z = S^1 \times S^1$. Then the Lie algebra $\underline{z}$ of Z is contained in the center of the Lie algebra of G and hence any one dimensional subspace $\mathfrak{h} \subset \underline{z}$ is an ideal in $\mathrm{GL}(G)$. To obtain the desired counterexample, take $\mathfrak{h}$ to be the Lie algebra corresponding to the subgroup $\{ (\mu, \nu) : \nu = \alpha\mu \} \subset S^1 \times S^1$, where α is irrational. This subgroup is connected and simply connected but not closed in G.

9. Cartan's theorems

Suppose $k = \mathbf{R}$ or $\mathbf{Q}_p$, that is, suppose $\mathbf{Q}$ is dense in k, $\operatorname{char} k = 0$.

Theorem 1. *Suppose G is an analytic subgroup over k and that $H \subset G$ is a topologically closed subgroup chunk. Then, H is analytic.*

Corollary. *A closed subgroup of an analytic group over* **R** *or* $\mathbf{Q}_p$ *is an analytic group.*

Theorem 2. *For $i = 1, 2$, let G_i be an analytic group over k. Then any continuous homomorphism $\phi : G_1 \to G_2$ is analytic.*

Proofs. 1. Theorem 1 $\Rightarrow$ Theorem 2:

Since ϕ is continuous, the graph $\Gamma_\phi \subset G_1 \times G_2$ of ϕ is a closed subgroup. Hence, by Theorem 1, Γ_ϕ is analytic. Let $p = \mathrm{pr}_1 |_{\Gamma_\phi}$. Then p is an analytic homomorphism with trivial kernel. Hence, $L(p)$ is injective and p is an immersion. Topologically, p is an isomorphism. It follows that p is an analytic isomorphism. Since $\phi = \mathrm{pr}_2 \, o p^{-1}$, ϕ is analytic.

2. Theorem 1 when $k = \mathbf{Q}_p$:

Let $\mathfrak{g} = L(G)$. Then, by taking a sufficiently small open subgroup of G, we may assume that G is isomorphic to an open subgroup U of $\mathfrak{g}$ under the Campbell-Hausdorff formula and that H is a closed subgroup of G (Chap. 4, §8, Cor. 2 of Thm.; Chap. 5, §7, Cor. of Thm. 1). We identify G with U. We then have that, for $x \in H$ and $n \in \mathbf{Z}$ $n \cdot x = f_n(x) \in H$. Since H is closed, we have the same statement for $n \in \mathbf{Z}_p$.

Choose $x_1, \ldots, x_m \in H$ such that $x_1, \ldots, x_m$ are linearly independent over $\mathbf{Q}_p$ and maximal with this property. Let V be the vector space generated by the $\{x_i\}$. Then, $X \subset V$, since otherwise $x_1, \ldots, x_m$ would not be a maximal linearly independent set in H. To prove the theorem, it suffices to show that H contains a neighborhood of 0 *in* V. Consider the map:

$$f : \mathbf{Z}_p^m \to V$$

defined by

$$f(t_1, \ldots, t_m) = (t_1 x_1) \cdots (t_m x_m) .$$

Then, f is analytic, and $Df(0)$ is bijective by construction; hence f is étale at 0. But $\mathrm{Im}(f) \subset H$ which shows that H contains a neighborhood of 0 in V.

3. Theorem 1 when $k = \mathbf{R}$.

Let $\mathfrak{g} = L(G)$. We may assume that H is a closed subgroup chunk of U where $U \subset \mathfrak{g}$ is an open subgroup chunk under the Campbell-Hausdorff formula. We may also assume that H is strict in U, that is:

a. $x, y \in H$ and $xy \in U \Rightarrow xy \in H$;

b. $x \in H \Rightarrow x^{-1} \in H$.

Let $V = \{ x \in \mathfrak{g} : tx \in H, \text{ for small } t \}$, that is, V consists of the points x in $\mathfrak{g}$ such that an interval about 0 on the ray through x lies in H. Then, we contend:

Lemma. 1. *V is a Lie subalgebra of $\mathfrak{g}$.*

2. *Suppose $x_n \in H$, $x_n \neq 0$. Let D_n be the line in $\mathfrak{g}$ containing x_n. Suppose $x_n \to 0$ and $D_n \to D$ as $n \to \infty$. Then, $D \subset V$.*

Proof. 2. Fix ε so that the ball of radius ε about 0 is contained in U. Let m be a positive integer and let $\varepsilon_m = \varepsilon / m$. Define:

$$S_i = \{ x : (i-1)\varepsilon_m \leq |x| \leq i\varepsilon_m \} .$$

In particular, S_1 is the ball of radius ε_m. For some constant K_m, $x_n \in S_1$ for all $n \geq K_m$. Consider any i such that $1 < i \leq m$. Then, for every $n \geq K_m$, there exists an integral multiple y_n^i of x_n lying in S_i. Since $D_n \to D$, as $n \to \infty$, a subsequence of y_n^i converges to a point in $S_i \cap D$. This point also lies in H since $y_n^i \in H$ by a) above and since H is closed. Hence, we have shown:

(∗) For any integer $m > 0$ and any integer i such that $1 < i \leq m$, there is an element $x \in H \cap D$ such that $(i-1)\varepsilon_m \leq |x| \leq i\varepsilon_m$.

Statement (∗) shows that H is dense in at least one of the two half intervals of length ε with endpoint 0 in D. By b) above, we see that H is dense in the symmetric interval of length 2ε about 0 in D and since H is closed we see that in fact $H \cap D$ contains this interval. This shows that $D \subset V$.

1. We use 2) to show that V is closed under addition and brackets. Let $x, y \in V$, $x, y \neq 0$. Then, by the Campbell-Hausdorff formula:

$$\lim_{n\to\infty} n\Big\{\Big(\frac{1}{n}x\Big)\cdot\Big(\frac{1}{n}y\Big)\Big\} = x+y$$

$$\lim_{n\to\infty} n^2\Big\{\Big[\Big(\frac{1}{n}x\Big),\Big(\frac{1}{n}y\Big)\Big]\Big\} = [x,y]$$

(See also Chap. 4, §7, n°5). The first formula shows that the line through $x+y$ satisfies the conditions of 2) while the second shows that the line through $[x,y]$ satisfies these conditions.

Since V is a Lie subalgebra of $\mathfrak{g}$, $V\cap U$ is an analytic subgroup chunk of U under the Campbell-Hausdorff formula. Using the assumptions of strictness of H, we see that $H\supset V\cap U$. The proof will therefore be complete if we show that H is contained in V in a neighborhood of 0. We suppose the contrary is true, that is, that there exists a sequence $\{x_n\}$ such that: $x_n \in H-V$, $x_n\to 0$ as $n\to\infty$. Choose a complement W of V in $\mathfrak{g}$. Then since exp is a local isomorphism at 0, we may write $x_n = w_n v_n$, $w_n\in W$ and $v_n\in V$, at least for $n \gg 0$. By strictness, $w_n \in H$ for $n\gg 0$. Hence, we can assume the original sequence $\{x_n\}$ belonged to W. Let D_n be the line through x_n. By the compactness of the projective space $\mathbf{P}(W)$, a subsequence of $\{D_n\}$ converges, say, to D. Then, by 2) of the Lemma, $D\subset V$ which is absurd.

Remark. Theorem 2 may be expressed by saying that the category of analytic groups over $k=\mathbf{R}$ or $\mathbf{Q}_p$ is a full subcategory of the category of all locally compact topological groups.

We may then ask: "When is a locally compact topological group a real or p-adic analytic group?" This question makes sense because Theorem 2 shows that if the structure of analytic group exists on a locally compact topological group, then it is unique.

The answers are:

1. Real case (Gleason-Montgomery-Zippin-Yamabe): The group G must contain no small subgroup (i.e., there is a neighborhood U of e such that any subgroup of G contained in U is equal to $\{e\}$).

2. p-adic case (Lazard): The group G must contain an open subgroup U with the following properties:

(a) U is a finitely generated pro-p-group.

(b) The commutator subgroup (U,U) is contained in U^{p^2} = set of p^2 powers.

In both cases, the *necessity* of the condition is easy (cf. Exer. 4).

Exercises

1. Let k be a field of char $p\neq 0$, let F be a formal group law over k, and let U (resp. $\mathfrak{g}$) be the corresponding bialgebra of point distributions (resp. the corresponding Lie algebra). One has $\mathfrak{g}\subset U$.

a) If $n = \dim \mathfrak{g}$, show that $\mathfrak{g}$ generates a subalgebra of U of rank p^n.

b) Show that $x \in \mathfrak{g} \Rightarrow x^p \in \mathfrak{g}$, where x^p denotes the pth-power of x in U. Show that $\operatorname{ad}(x^p) = \operatorname{ad}(x)^p$.

c) Let a be an element of k which does not belong to the prime field $\mathbf{F}_p$. Let $\mathfrak{h}$ be the Lie algebra with basis $\{X, Y, Z\}$ and relations $[X, Y] = Y$, $[X, Z] = aZ$, $[Y, Z] = 0$. Show that there is no element $y \in \mathfrak{h}$ such that $\operatorname{ad}(y) = \operatorname{ad}(X)^p$. Prove that $\mathfrak{h}$ cannot be the Lie algebra of a formal group.

2. Let $H_1 = k[[X]]$ and $H_2 = k[[Y]]$.

a) Suppose k is a field. Show that any algebra homomorphism $\phi : H_2 \to H_1$ is admissible (cf. §6).

b) Suppose k has no nilpotent elements (except 0). Show that any continuous algebra homomorphism $\phi : H_2 \to H_1$ is admissible.

3. Let $k = \mathbf{R}$ or $\mathbf{C}$, and let s be a semisimple subalgebra of the Lie algebra of $\mathrm{GL}(n, k)$. Show that s corresponds to a group submanifold of $\mathrm{GL}(n, k)$. (Hint: use L.A., Chap. 6, Theorem 5.2.)

4. Let G be a standard p-adic group (cf. Chap. 4), and let $\{G_n\}$ be its canonical filtration. Show that, if $U = G_n$ with $n \geq 2$, one has:

$$(U, U) \subset U^{p^n} .$$

5. Let G be a real Lie group, with Lie algebra $\mathfrak{g}$, let $\mathfrak{h}$ be a subalgebra of $\mathfrak{g}$, and let H be the Lie subgroup of G corresponding to $\mathfrak{h}$. Assume that H is dense in G.

a) Show that $\operatorname{Ad}(g)\mathfrak{h} = \mathfrak{h}$ for all $g \in G$, and that $\mathfrak{h}$ is an ideal of $\mathfrak{g}$.

b) Let $\tilde{G}$ be the universal covering of G, let Z be the kernel of $\tilde{G} \to G$, and let $\tilde{H}$ be the Lie subgroup of $\tilde{G}$ corresponding to $\mathfrak{h}$; $\tilde{H}$ is closed in $\tilde{G}$ (§8, Theorem 4). Show that $\tilde{H} \cdot Z$ is dense in $\tilde{G}$, and that $\tilde{G}/\tilde{H}$ is abelian, hence that $\mathfrak{g}/\mathfrak{h}$ is abelian.

c) Suppose $\mathfrak{g}$ is semisimple. Show that $\tilde{G} = \tilde{H} \times \mathbf{R}^n$ for some n. Show that $n = 0$ (hence $G = H$) if the center of H is finite.

d) Let $H_0 = \mathrm{SL}(2, \mathbf{R})$. Show that $\pi_1(H_0) = \mathbf{Z}$. Show that the universal covering H of H_0 can be imbedded as a dense Lie subgroup in a Lie group G of arbitrary dimension ≥ 3.

6. Let G be a real Lie group, with Lie algebra $\mathfrak{g}$. For any subalgebra $\mathfrak{h}$ of $\mathfrak{g}$, let H be the corresponding Lie subgroup of G. The adherence $\overline{H}$ of H is a closed Lie subgroup of G (by Cartan's theorem); let $\overline{\mathfrak{h}}$ be its Lie algebra.

a) Show that $\mathfrak{h} \subset \overline{\mathfrak{h}}$, $\overline{\overline{\mathfrak{h}}} = \overline{\mathfrak{h}}$, $\overline{\mathfrak{h}_1 \cap \mathfrak{h}_2} \subset \overline{\mathfrak{h}_1} \cap \overline{\mathfrak{h}_2}$

b) Show that $\mathfrak{h}$ is an ideal in $\overline{\mathfrak{h}}$, and that $\overline{\mathfrak{h}}/\mathfrak{h}$ is abelian (use Exer. 5).

Appendix. Existence theorem for ordinary differential equations

We assume $\operatorname{char} k = 0$.

Theorem. *Suppose $\phi = (\phi_1, \ldots, \phi_n)$ is a system of n convergent power series in n variables. Then:*

1. *The formal differential equation*

$$\tau'(s) = \phi(\tau(r)) \qquad \textit{and} \qquad \tau(0) = 0 \ ,$$

possesses a unique solution τ.

2. τ *is convergent.*

Proof.

Case 1: $k = \mathbf{R}$ or $\mathbf{C}$.

1. Write:

$$\tau(s) = \sum_{n\geq 1} a_n s^n \ ;$$
$$\phi(X) = \sum c_\alpha X^\alpha \ .$$

Then, the formal differential equation takes the form:

$$\sum_{n\geq 1} n a_n s^{n-1} = \sum c_\alpha \Big(\sum_{m\geq 1} a_m s^m \Big)^\alpha \ .$$

Then, there exist unique polynomials $Q_n(c_\alpha, a_m)$, $|\alpha|$, $m < n$, with positive integral coefficients such that:

$$a_n = \frac{1}{n} Q_n(c_\alpha, a_m) \ .$$

This shows the uniqueness of the formal solution, by induction on n.

2. To show convergence, we use Cauchy's method of majorants. Suppose $|c_\alpha| \leq d_\alpha$ where $\{d_\alpha\}$ consists of non-negative real numbers. Let $\bar{\tau}(s) = \sum b_n t^n$ be the formal solution of the differential equation corresponding to $\bar{\phi}(X) = \sum d_\alpha X^\alpha$. Then:

Lemma 1. $\bar{\tau}$ *convergent* $\Rightarrow$ τ *convergent. More precisely,* $\bar{\tau}$ *is a majorant for* τ.

Proof. By induction:

$$|a_n| = \left|\frac{1}{n}\right| |Q_n(c_\alpha, a_m)| \leq \frac{1}{n} Q_n(|c_\alpha|, |a_m|) \leq \frac{1}{n} Q_n(d_\alpha, b_m) = b_n \ .$$

Note that we have used the fact that $k = \mathbf{R}$ or $\mathbf{C}$ to obtain the equality:

$$\left|\frac{1}{n}\right| = \frac{1}{n} \ .$$

To apply Lemma 1, we must construct an appropriate $\bar{\phi}$ and compute the corresponding $\bar{\tau}$ explicitly. Since ϕ is convergent, we may find constants $M, R > 0$ such that:

$$\sum |c_\alpha| R^{|\alpha|} < M \ .$$

Let $d_\alpha = \frac{M}{R^{|\alpha|}}$. Clearly $|c_\alpha| \leq d_\alpha$, and we have that:

$$\bar{\phi}(X) = \sum M \cdot \left(\frac{X}{R}\right)^\alpha = \frac{M}{\prod_i (1 - X_i/R)} .$$

By the uniqueness statement, $\bar{\tau}(s) = (\sigma(s), \ldots, \sigma(s))$ where $\sigma(s)$ is the formal solution of the single differential equation:

$$\sigma'(s) = \frac{M}{(1 - \frac{\sigma(s)}{R})^n} .$$

We make an explicit computation for $\sigma(s)$ which shows that $\sigma(s)$ is convergent:

$$\sigma(s) = R(1 - \left\{1 - (n+1)M \cdot \frac{s}{R}\right\}^{\frac{1}{n+1}}) .$$

Indeed:

$$1 - \frac{\sigma(s)}{R} = \left\{1 - (n+1)M \cdot \frac{s}{R}\right\}^{\frac{1}{n+1}} .$$

Differentiating $\sigma(s)$ and using the above formula, one sees that $\sigma(s)$ does satisfy the desired differential equation.

Case 2: k ultrametric.

Since ϕ is convergent, we may assume, by change of coordinates *via* a homothety, that the coefficients of ϕ lie in the valuation ring A of k.

1. Write:

$$\tau(s) = \sum_{n \geq 1} a_n \frac{s^n}{n!}$$

$$\phi(X) = \sum c_\alpha X^\alpha .$$

Then, the formal differential equation takes the form:

$$\sum_{n \geq 0} a_{n+1} \frac{s^n}{n!} = \sum c_\alpha \left(\sum_{m \geq 1} a_m \frac{s^m}{m!}\right)^\alpha .$$

Then, using the fact the binomial coefficients lie in $\mathbf{Z}$, we see that there exist unique polynomials $Q_n(c_\alpha, a_m)$, $|\alpha|$, $m < n$, with positive integral coefficients such that:

$$a_n = Q_n(c_\alpha, a_m) .$$

This shows the uniqueness of the formal solution.

2. By induction on n, $a_n \in A$ since by assumption all $c_\alpha \in A$. Hence, by Lemma 4 of §4, for some real constant a, $0 < a \leq 1$, $\tau(s)$ is majorized by:

$$\sum \frac{r^n}{a^n} .$$

Since this is a geometric series, it converges for small r, so that τ is convergent.

Bibliography

Lie Algebras

Bourbaki, N.: Groupes et Algèbres de Lie, Chap. 1, 2, 7, 8. Hermann, Paris (English translation: Springer-Verlag).

Chevalley, C.: Théorie des groupes de Lie, tome III. Publ. Inst. Math. Nancago IV, Hermann, Paris, 1955.

Humphreys, J.: Introduction to Lie Algebras and Representation Theory. GTM 9, Springer-Verlag, Heidelberg, 1972.

Jacobson, N.: Lie Algebras. Intersc. Tracts, n°10, John Wiley and Sons, New York, 1962.

Séminaire Sophus Lie. Théorie des algèbres de Lie – Topologie des groupes de Lie. Secr. Math., rue P. Curie, Paris, 1955.

Serre, J.-P.: Algèbres de Lie semi-simples complexes. Benjamin, New York, 1966 (English translation: Springer-Verlag).

Formal Groups

Dieudonné, J.: Introduction to the theory of formal groups. Marcel Dekker, Inc., New York, 1973.

Fröhlich, A.: Formal groups. Lect. Notes in Math., **74**, Springer-Verlag, Berlin, 1968.

Lazard, M.: Sur les groupes de Lie formels à un paramètre. Bull. Soc. Math. France, **83**, 1955, p. 251–274.

Lazard, M.: Lois des groupes et analyseurs. Ann. ENS, **72**, 1955, p. 299–400.

Manin, Yu.: The theory of commutative formal groups over fields of finite characteristic. Usp. Mat. Nauk, **18**, 1963, p. 3–91 (Russ. Math. Surveys, **18**, 1963, p. 1–84).

Differentiable Manifolds

Bourbaki, N.: Variétés différentielles et analytiques. Fasc. de Rés., §§1–7 and §§8–15, Hermann, Paris, 1971.

Dieudonné, J.: Eléments d'Analyse (tome 3). Gauthier-Villars, Paris, 1970.

Lang, S.: Differential Manifolds. Addison-Wesley, Reading, 1972.

Warner, F.: Foundations of Differentiable Manifolds and Lie Groups. Scott, Foresman, Glenview, Illinois, 1971.

Topological Groups

Montgomery, D. and Zippin, L.: Topological transformation groups. Intersc., New York, 1955.

Pontrjagin, L.: Topological Groups. Univ. Press, Princeton, 1939.

Lie Groups:

Bourbaki, N.: Groupes et Algèbres de Lie, Chap. 3. Hermann, Paris (English translation: Springer-Verlag).

Chevalley, C.: Theory of Lie groups. Univ. Press, Princeton, 1946.

Helgason, S.: Differential Geometry and Symmetric Spaces. Acad. Press, New York, 1962.

Hochschild, G.: The structure of Lie groups. Holden-Day, San Francisco, 1965.

p-adic groups
Lazard, M.: Groupes analytiques *p*-adiques. Publ. Math. IHES, **26**, PUF, Paris, 1965.

Algebraic Groups:
Borel, A.: Linear Algebraic Groups, 2nd edit., Springer-Verlag, 1991
Chevalley, C.: Sur certains groupes simples. Tôh. Math. J., **7**, 1955, p. 14–66.
Chevalley, C.: Classification des groupes de Lie algébriques. Secr. Math., IHP, rue P. Curie, Paris, 1958.
Demazure, M. et Gabriel, P.: Groupes algébriques (tome I). Masson, Paris, 1970.

Problem

(Harvard Exam., Jan. 1965 – Time: 3 hours)

In what follows k denotes a field, and $\mathfrak{g}$ a 3-dimensional Lie algebra over k, with basis $\{x, y, z\}$ and relations:

$$[x, y] = z \ , \quad [x, z] = [y, z] = 0 \ .$$

The universal algebra $U\mathfrak{g}$ of $\mathfrak{g}$ is denoted by U.

I

1. Determine the center of $\mathfrak{g}$. Prove that $\mathfrak{g}$ is nilpotent.

2. Let A be the center of U. Show that $z \in A$. If k is of characteristic $p \neq 0$, show that A also contains x^p and y^p, and that z, x^p, y^p are algebraically independent.

3. Give an example of an analytic group (over some complete field k) having a Lie algebra isomorphic to $\mathfrak{g}$.

II

In this section V is a vector space over k, and $\varrho : \mathfrak{g} \to \mathrm{End}(V)$ is a Lie algebra homomorphism (so that V is a $\mathfrak{g}$-module).

4. For any $\lambda \in k$, let V_λ be the set of $v \in V$ such that $\varrho(z)v = \lambda v$. Show that V_λ is a $\mathfrak{g}$-submodule of V.

5. Assume k algebraically closed, and V irreducible(*) of finite dimension. Show that there exists $\lambda \in k$ such that $\varrho(z) = \lambda$, scalar multiplication by λ. Assume moreover that $\mathrm{char}(k) = 0$; show that $\lambda = 0$ and classify all irreducible $\mathfrak{g}$-modules of finite dimension.

6. We now take for V the vector space $k[T]$ of polynomials in one indeterminate T. Show that there exists a structure of $\mathfrak{g}$-module on V such that, if $P \in k[T]$:

$$\varrho(x) \cdot P = dP(T)/dT \ , \quad \varrho(y) \cdot P = T \cdot P(T) \ , \quad \varrho(z) \cdot P = P \ .$$

Prove that V is irreducible if $\mathrm{char}(k) = 0$.

(*) A $\mathfrak{g}$-module V is said to be irreducible if $V \neq 0$ and if the only $\mathfrak{g}$-submodules of V are 0 and V.

III

In this section, k is algebraically closed of char. $p \neq 0$.

7. Let V be the $\mathfrak{g}$-module defined in question 6. Show that the $\mathfrak{g}$-submodules of V are of the form $V_P = P(T^p) \cdot V$, with $P \in k[T]$. Show that V/V_P is irreducible if and only if $\deg(P) = 1$.

8. Let W be an irreducible $\mathfrak{g}$-module, and let $\varrho_W : \mathfrak{g} \to \operatorname{End}(W)$ be the corresponding homomorphism. Show that W is isomorphic to one of the modules V/V_P defined above if and only if the following two conditions are satisfied: $\varrho_W(z) = 1$, and $\varrho_W(x)$ is nilpotent.

9. Let again W be an irreducible $\mathfrak{g}$-module of finite dimension, and assume $\dim(W) > 1$. Show that $\dim(W) = p$, that $\varrho_W(z)$ is equal to a scalar $\lambda \neq 0$, that $\varrho_W(x)$ has only one eigenvalue μ, and that $\varrho_W(y)$ has only one eigenvalue ν. Show that, for any (λ, μ, ν) with $\lambda \neq 0$, there exists a corresponding W, and that it is unique, up to isomorphism.

10. Prove that the center A of U is the polynomial algebra generated by z, x^p, y^p. If k' is an extension of k, and $\varphi : A \to k'$ any homomorphism such that $\varphi(z) \neq 0$, show that $U \otimes_k k'$ is a central simple algebra over k' of rank p^2. Prove that this remains true even if k is not algebraically closed.

11. Prove that every irreducible $\mathfrak{g}$-module is finite-dimensional.

INDEX

Lecture Notes in Mathematics

For information about Vols. 1–1312
please contact your bookseller or Springer-Verlag

Vol. 1313: F. Colonius. Optimal Periodic Control. VI, 177 pages. 1988.

Vol. 1314: A. Futaki. Kähler-Einstein Metrics and Integral Invariants. IV, 140 pages. 1988.

Vol. 1315: R.A. McCoy, I. Ntantu, Topological Properties of Spaces of Continuous, Functions. IV, 124 pages. 1988.

Vol. 1316: H. Korezlioglu, A.S. Ustunel (Eds.), Stochastic Analysis and Related Topics. Proceedings, 1986. V, 371 pages. 1988.

Vol. 1317: J. Lindenstrauss, V.D. Milman (Eds.), Geometric Aspects of Functional Analysis. Seminar, 1986–87. VII, 289 pages. 1988.

Vol. 1318: Y. Felix (Ed.), Algebraic Topology – Rational Homotopy. Proceedings, 1986. VIII, 245 pages. 1988.

Vol. 1319: M. Vuorinen, Conformal Geometry and Quasiregular Mappings. XIX, 209 pages. 1988.

Vol. 1320: H. Jürgensen, G. Lallement, H.J. Weinert (Eds.), Semigroups, Theory and Applications. Proceedings, 1986. X, 416 pages. 1988.

Vol. 1321: J. Azéma, P.A. Meyer, M. Yor (Eds.), Séminaire de Probabilités XXII. Proceedings. IV, 600 pages. 1988.

Vol. 1322: M. Métivier, S. Watanabe (Eds.), Stochastic Analysis. Proceedings, 1987. VII, 197 pages. 1988.

Vol. 1323: D.R. Anderson, H.J. Munkholm, Boundedly Controlled Topology. XII, 309 pages. 1988.

Vol. 1324: F. Cardoso, D.G. de Figueiredo, R. Iório, O. Lopes (Eds.), Partial Differential Equations. Proceedings, 1986. VIII, 433 pages. 1988.

Vol. 1325: A. Truman, I.M. Davies (Eds.), Stochastic Mechanics and Stochastic Processes. Proceedings, 1986. V, 220 pages. 1988.

Vol. 1326: P.S. Landweber (Ed.), Elliptic Curves and Modular Forms in Algebraic Topology. Proceedings, 1986. V, 224 pages. 1988.

Vol. 1327: W. Bruns, U. Vetter, Determinantal Rings. VII, 236 pages. 1988.

Vol. 1328: J.L. Bueso, P. Jara, B. Torrecillas (Eds.), Ring Theory. Proceedings, 1986. IX, 331 pages. 1988.

Vol. 1329: M. Alfaro, J.S. Dehesa, F.J. Marcellan, J.L. Rubio de Francia, J. Vinuesa (Eds.): Orthogonal Polynomials and their Applications. Proceedings, 1986. XV, 334 pages. 1988.

Vol. 1330: A. Ambrosetti, F. Gori, R. Lucchetti (Eds.), Mathematical Economics. Montecatini Terme 1986. Seminar. VII, 137 pages. 1988.

Vol. 1331: R. Bamón, R. Labarca, J. Palis Jr. (Eds.), Dynamical Systems, Valparaiso 1986. Proceedings. VI, 250 pages. 1988.

Vol. 1332: E. Odell, H. Rosenthal (Eds.), Functional Analysis. Proceedings. 1986–87. V, 202 pages. 1988.

Vol. 1333: A.S. Kechris, D.A. Martin, J.R. Steel (Eds.), Cabal Seminar 81–85. Proceedings, 1981–85. V, 224 pages. 1988.

Vol. 1334: Yu.G. Borisovich, Yu.E. Gliklikh (Eds.), Global Analysis – Studies and Applications III. V, 331 pages. 1988.

Vol. 1335: F. Guillén, V. Navarro Aznar, P. Pascual-Gainza, F. Puerta, Hyperrésolutions cubiques et descente cohomologique. XII, 192 pages. 1988.

Vol. 1336: B. Helffer, Semi-Classical Analysis for the Schrödinger Operator and Applications. V, 107 pages. 1988.

Vol. 1337: E. Sernesi (Ed.), Theory of Moduli. Seminar, 1985. VIII, 232 pages. 1988.

Vol. 1338: A.B. Mingarelli, S.G. Halvorsen. Non-Oscillation Domains of Differential Equations with Two Parameters. XI, 109 pages. 1988.

Vol. 1339: T. Sunada (Ed.), Geometry and Analysis of Manifolds. Proceedings, 1987. IX, 277 pages. 1988.

Vol. 1340: S. Hildebrandt, D.S. Kinderlehrer, M. Miranda (Eds.), Calculus of Variations and Partial Differential Equations. Proceedings, 1986. IX, 301 pages. 1988.

Vol. 1341: M. Dauge, Elliptic Boundary Value Problems on Corner Domains. VIII, 259 pages. 1988.

Vol. 1342: J.C. Alexander (Ed.), Dynamical Systems. Proceedings, 1986–87. VIII, 726 pages. 1988.

Vol. 1343: H. Ulrich, Fixed Point Theory of Parametrized Equivariant Maps. VII, 147 pages. 1988.

Vol. 1344: J. Král, J. Lukes, J. Netuka, J. Vesely´ (Eds.), Potential Theory – Surveys and Problems. Proceedings, 1987. VIII, 271 pages. 1988.

Vol. 1345: X. Gomez-Mont, J. Seade, A. Verjovski (Eds.), Holomorphic Dynamics. Proceedings, 1986. VII. 321 pages. 1988.

Vol. 1346: O.Ya. Viro (Ed.), Topology and Geometry – Rohlin Seminar. XI, 581 pages. 1988.

Vol. 1347: C. Preston, Iterates of Piecewise Monotone Mappings on an Interval. V, 166 pages. 1988.

Vol. 1348: F. Borceux (Ed.), Categorical Algebra and its Applications. Proceedings, 1987. VIII, 375 pages. 1988.

Vol. 1349: E. Novak, Deterministic and Stochastic Error Bounds in Numerical Analysis. V, 113 pages. 1988.

Vol. 1350: U. Koschorke (Ed.), Differential Topology Proceedings, 1987, VI, 269 pages. 1988.

Vol. 1351: I. Laine, S. Rickman, T. Sorvali (Eds.), Complex Analysis, Joensuu 1987. Proceedings. XV, 378 pages. 1988.

Vol. 1352: L.L. Avramov, K.B. Tchakerian (Eds.), Algebra – Some Current Trends. Proceedings. 1986. IX, 240 Seiten. 1988.

Vol. 1353: R.S. Palais, Ch.-l. Teng, Critical Point Theory and Submanifold Geometry. X, 272 pages. 1988.

Vol. 1354: A. Gómez, F. Guerra, M.A. Jiménez, G. López (Eds.), Approximation and Optimization. Proceedings, 1987. VI, 280 pages. 1988.

Vol. 1355: J. Bokowski, B. Sturmfels, Computational Synthetic Geometry. V, 168 pages. 1989.

Vol. 1356: H. Volkmer, Multiparameter Eigenvalue Problems and Expansion Theorems. VI, 157 pages. 1988.

Vol. 1357: S. Hildebrandt, R. Leis (Eds.), Partial Differential Equations and Calculus of Variations. VI, 423 pages. 1988.

Vol. 1358: D. Mumford, The Red Book of Varieties and Schemes. V, 309 pages. 1988.

Vol. 1359: P. Eymard, J.-P. Pier (Eds.) Harmonic Analysis. Proceedings, 1987. VIII, 287 pages. 1988.

Vol. 1360: G. Anderson, C. Greengard (Eds.), Vortex Methods. Proceedings, 1987. V, 141 pages. 1988.

Vol. 1361: T. tom Dieck (Ed.), Algebraic Topology and Transformation Groups. Proceedings. 1987. VI, 298 pages. 1988.

Vol. 1362: P. Diaconis, D. Elworthy, H. Föllmer, E. Nelson, G.C. Papanicolaou, S.R.S. Varadhan. École d´ Été de Probabilités de Saint-Flour XV–XVII. 1985–87 Editor: P.L. Hennequin. V, 459 pages. 1988.

Vol. 1363: P.G. Casazza, T.J. Shura, Tsirelson´s Space. VIII, 204 pages. 1988.

Vol. 1364: R.R. Phelps, Convex Functions, Monotone Operators and Differentiability. IX, 115 pages. 1989.

Vol. 1365: M. Giaquinta (Ed.), Topics in Calculus of Variations. Seminar, 1987. X, 196 pages. 1989.

Vol. 1366: N. Levitt, Grassmannians and Gauss Maps in PL-Topology. V, 203 pages. 1989.

Vol. 1367: M. Knebusch, Weakly Semialgebraic Spaces. XX, 376 pages. 1989.

Vol. 1368: R. Hübl, Traces of Differential Forms and Hochschild Homology. III, 111 pages. 1989.

Vol. 1369: B. Jiang, Ch.-K. Peng, Z. Hou (Eds.), Differential Geometry and Topology. Proceedings, 1986–87. VI, 366 pages. 1989.

Vol. 1370: G. Carlsson, R.L. Cohen, H.R. Miller, D.C. Ravenel (Eds.), Algebraic Topology. Proceedings, 1986. IX, 456 pages. 1989.

Vol. 1371: S. Glaz, Commutative Coherent Rings. XI, 347 pages. 1989.

Vol. 1372: J. Azéma, P.A. Meyer, M. Yor (Eds.), Séminaire de Probabilités XXIII. Proceedings. IV, 583 pages. 1989.

Vol. 1373: G. Benkart, J.M. Osborn (Eds.), Lie Algebras. Madison 1987. Proceedings. V, 145 pages. 1989.

Vol. 1374: R.C. Kirby, The Topology of 4-Manifolds. VI, 108 pages. 1989.

Vol. 1375: K. Kawakubo (Ed.), Transformation Groups. Proceedings, 1987. VIII, 394 pages, 1989.

Vol. 1376: J. Lindenstrauss, V.D. Milman (Eds.), Geometric Aspects of Functional Analysis. Seminar (GAFA) 1987–88. VII, 288 pages. 1989.

Vol. 1377: J.F. Pierce, Singularity Theory, Rod Theory, and Symmetry-Breaking Loads. IV, 177 pages. 1989.

Vol. 1378: R.S. Rumely, Capacity Theory on Algebraic Curves. III, 437 pages. 1989.

Vol. 1379: H. Heyer (Ed.), Probability Measures on Groups IX. Proceedings, 1988. VIII, 437 pages. 1989.

Vol. 1380: H.P. Schlickewei, E. Wirsing (Eds.), Number Theory, Ulm 1987. Proceedings. V, 266 pages. 1989.

Vol. 1381: J.-O. Strömberg, A. Torchinsky, Weighted Hardy Spaces. V, 193 pages. 1989.

Vol. 1382: H. Reiter, Metaplectic Groups and Segal Algebras. XI, 128 pages. 1989.

Vol. 1383: D.V. Chudnovsky, G.V. Chudnovsky, H. Cohn, M.B. Nathanson (Eds.), Number Theory, New York 1985–88. Seminar. V, 256 pages. 1989.

Vol. 1384: J. Garcia-Cuerva (Ed.), Harmonic Analysis and Partial Differential Equations. Proceedings, 1987. VII, 213 pages. 1989.

Vol. 1385: A.M. Anile, Y. Choquet-Bruhat (Eds.), Relativistic Fluid Dynamics. Seminar, 1987. V, 308 pages. 1989.

Vol. 1386: A. Bellen, C.W. Gear, E. Russo (Eds.), Numerical Methods for Ordinary Differential Equations. Proceedings, 1987. VII, 136 pages. 1989.

Vol. 1387: M. Petkovi´c, Iterative Methods for Simultaneous Inclusion of Polynomial Zeros. X, 263 pages. 1989.

Vol. 1388: J. Shinoda, T.A. Slaman, T. Tugué (Eds.), Mathematical Logic and Applications. Proceedings, 1987. V, 223 pages. 1989.

Vol. 1000: Second Edition. H. Hopf, Differential Geometry in the Large. VII, 184 pages. 1989.

Vol. 1389: E. Ballico, C. Ciliberto (Eds.), Algebraic Curves and Projective Geometry. Proceedings, 1988. V, 288 pages. 1989.

Vol. 1390: G. Da Prato, L. Tubaro (Eds.), Stochastic Partial Differential Equations and Applications II. Proceedings, 1988. VI, 258 pages. 1989.

Vol. 1391: S. Cambanis, A. Weron (Eds.), Probability Theory on Vector Spaces IV. Proceedings, 1987. VIII, 424 pages. 1989.

Vol. 1392: R. Silhol, Real Algebraic Surfaces. X, 215 pages. 1989.

Vol. 1393: N. Bouleau, D. Feyel, F. Hirsch, G. Mokobodzki (Eds.), Séminaire de Théorie du Potentiel Paris, No. 9. Proceedings. VI, 265 pages. 1989.

Vol. 1394: T.L. Gill, W.W. Zachary (Eds.), Nonlinear Semigroups, Partial Differential Equations and Attractors. Proceedings, 1987. IX, 233 pages. 1989.

Vol. 1395: K. Alladi (Ed.), Number Theory, Madras 1987. Proceedings. VII, 234 pages. 1989.

Vol. 1396: L. Accardi, W. von Waldenfels (Eds.), Quantum Probability and Applications IV. Proceedings, 1987. VI, 355 pages. 1989.

Vol. 1397: P.R. Turner (Ed.), Numerical Analysis and Parallel Processing. Seminar, 1987. VI, 264 pages. 1989.

Vol. 1398: A.C. Kim, B.H. Neumann (Eds.), Groups – Korea 1988. Proceedings. V, 189 pages. 1989.

Vol. 1399: W.-P. Barth, H. Lange (Eds.), Arithmetic of Complex Manifolds. Proceedings, 1988. V, 171 pages. 1989.

Vol. 1400: U. Jannsen. Mixed Motives and Algebraic K-Theory. XIII, 246 pages. 1990.

Vol. 1401: J. Steprans, S. Watson (Eds.), Set Theory and its Applications. Proceedings, 1987. V, 227 pages. 1989.

Vol. 1402: C. Carasso, P. Charrier, B. Hanouzet, J.-L. Joly (Eds.), Nonlinear Hyperbolic Problems. Proceedings, 1988. V, 249 pages. 1989.

Vol. 1403: B. Simeone (Ed.), Combinatorial Optimization. Seminar, 1986. V, 314 pages. 1989.

Vol. 1404: M.-P. Malliavin (Ed.), Séminaire d´Algèbre Paul Dubreil et Marie-Paul Malliavin. Proceedings, 1987–1988. IV, 410 pages. 1989.

Vol. 1405: S. Dolecki (Ed.), Optimization. Proceedings, 1988. V, 223 pages. 1989. Vol. 1406: L. Jacobsen (Ed.), Analytic Theory of Continued Fractions III. Proceedings, 1988. VI, 142 pages. 1989.

Vol. 1407: W. Pohlers, Proof Theory. VI, 213 pages. 1989.

Vol. 1408: W. Lück, Transformation Groups and Algebraic K-Theory. XII, 443 pages. 1989.

Vol. 1409: E. Hairer, Ch. Lubich, M. Roche. The Numerical Solution of Differential-Algebraic Systems by Runge-Kutta Methods. VII, 139 pages. 1989.

Vol. 1410: F.J. Carreras, O. Gil-Medrano, A.M. Naveira (Eds.), Differential Geometry. Proceedings, 1988. V, 308 pages. 1989.

Vol. 1411: B. Jiang (Ed.), Topological Fixed Point Theory and Applications. Proceedings. 1988. VI, 203 pages. 1989.

Vol. 1412: V.V. Kalashnikov, V.M. Zolotarev (Eds.), Stability Problems for Stochastic Models. Proceedings, 1987. X, 380 pages. 1989.

Vol. 1413: S. Wright, Uniqueness of the Injective III_1 Factor. III, 108 pages. 1989.

Vol. 1414: E. Ramirez de Arellano (Ed.), Algebraic Geometry and Complex Analysis. Proceedings, 1987. VI, 180 pages. 1989.

Vol. 1415: M. Langevin, M. Waldschmidt (Eds.), Cinquante Ans de Polynômes. Fifty Years of Polynomials. Proceedings, 1988. IX, 235 pages.1990.

Vol. 1416: C. Albert (Ed.), Géométrie Symplectique et Mécanique. Proceedings, 1988. V, 289 pages. 1990.

Vol. 1417: A.J. Sommese, A. Biancofiore, E.L. Livorni (Eds.), Algebraic Geometry. Proceedings, 1988. V, 320 pages. 1990.

Vol. 1418: M. Mimura (Ed.), Homotopy Theory and Related Topics. Proceedings, 1988. V, 241 pages. 1990.

Vol. 1419: P.S. Bullen, P.Y. Lee, J.L. Mawhin, P. Muldowney, W.F. Pfeffer (Eds.), New Integrals. Proceedings, 1988. V, 202 pages. 1990.

Vol. 1420: M. Galbiati, A. Tognoli (Eds.), Real Analytic Geometry. Proceedings, 1988. IV, 366 pages. 1990.

Vol. 1421: H.A. Biagioni, A Nonlinear Theory of Generalized Functions, XII, 214 pages. 1990.

Vol. 1422: V. Villani (Ed.), Complex Geometry and Analysis. Proceedings, 1988. V, 109 pages. 1990.

Vol. 1423: S.O. Kochman, Stable Homotopy Groups of Spheres: A Computer-Assisted Approach. VIII, 330 pages. 1990.

Vol. 1424: F.E. Burstall, J.H. Rawnsley, Twistor Theory for Riemannian Symmetric Spaces. III, 112 pages. 1990.

Vol. 1425: R.A. Piccinini (Ed.), Groups of Self-Equivalences and Related Topics. Proceedings, 1988. V, 214 pages. 1990.

Vol. 1426: J. Azéma, P.A. Meyer, M. Yor (Eds.), Séminaire de Probabilités XXIV, 1988/89. V, 490 pages. 1990.

Vol. 1427: A. Ancona, D. Geman, N. Ikeda, École d'Eté de Probabilités de Saint Flour XVIII, 1988. Ed.: P.L. Hennequin. VII, 330 pages. 1990.

Vol. 1428: K. Erdmann, Blocks of Tame Representation Type and Related Algebras. XV. 312 pages. 1990.

Vol. 1429: S. Homer, A. Nerode, R.A. Platek, G.E. Sacks, A. Scedrov, Logic and Computer Science. Seminar, 1988. Editor: P. Odifreddi. V, 162 pages. 1990.

Vol. 1430: W. Bruns, A. Simis (Eds.), Commutative Algebra. Proceedings. 1988. V, 160 pages. 1990.

Vol. 1431: J.G. Heywood, K. Masuda, R. Rautmann, V.A. Solonnikov (Eds.), The Navier-Stokes Equations – Theory and Numerical Methods. Proceedings, 1988. VII, 238 pages. 1990.

Vol. 1432: K. Ambos-Spies, G.H. Müller, G.E. Sacks (Eds.), Recursion Theory Week. Proceedings, 1989. VI, 393 pages. 1990.

Vol. 1433: S. Lang, W. Cherry, Topics in Nevanlinna Theory. II, 174 pages.1990.

Vol. 1434: K. Nagasaka, E. Fouvry (Eds.), Analytic Number Theory. Proceedings, 1988. VI, 218 pages. 1990.

Vol. 1435: St. Ruscheweyh, E.B. Saff, L.C. Salinas, R.S. Varga (Eds.), Computational Methods and Function Theory. Proceedings, 1989. VI, 211 pages. 1990.

Vol. 1436: S. Xambó-Descamps (Ed.), Enumerative Geometry. Proceedings, 1987. V, 303 pages. 1990.

Vol. 1437: H. Inassaridze (Ed.), K-theory and Homological Algebra. Seminar, 1987–88. V, 313 pages. 1990.

Vol. 1438: P.G. Lemarié (Ed.) Les Ondelettes en 1989. Seminar. IV, 212 pages. 1990.

Vol. 1439: E. Bujalance, J.J. Etayo, J.M. Gamboa, G. Gromadzki. Automorphism Groups of Compact Bordered Klein Surfaces: A Combinatorial Approach. XIII, 201 pages. 1990.

Vol. 1440: P. Latiolais (Ed.), Topology and Combinatorial Groups Theory. Seminar, 1985–1988. VI, 207 pages. 1990.

Vol. 1441: M. Coornaert, T. Delzant, A. Papadopoulos. Géométrie et théorie des groupes. X, 165 pages. 1990.

Vol. 1442: L. Accardi, M. von Waldenfels (Eds.), Quantum Probability and Applications V. Proceedings, 1988. VI, 413 pages. 1990.

Vol. 1443: K.H. Dovermann, R. Schultz, Equivariant Surgery Theories and Their Periodicity Properties. VI, 227 pages. 1990.

Vol. 1444: H. Korezlioglu, A.S. Ustunel (Eds.), Stochastic Analysis and Related Topics VI. Proceedings, 1988. V, 268 pages. 1990.

Vol. 1445: F. Schulz, Regularity Theory for Quasilinear Elliptic Systems and – Monge Ampère Equations in Two Dimensions. XV, 123 pages. 1990.

Vol. 1446: Methods of Nonconvex Analysis. Seminar, 1989. Editor: A. Cellina. V, 206 pages. 1990.

Vol. 1447: J.-G. Labesse, J. Schwermer (Eds), Cohomology of Arithmetic Groups and Automorphic Forms. Proceedings, 1989. V, 358 pages. 1990.

Vol. 1448: S.K. Jain, S.R. López-Permouth (Eds.), Non-Commutative Ring Theory. Proceedings, 1989. V, 166 pages. 1990.

Vol. 1449: W. Odyniec, G. Lewicki, Minimal Projections in Banach Spaces. VIII, 168 pages. 1990.

Vol. 1450: H. Fujita, T. Ikebe, S.T. Kuroda (Eds.), Functional-Analytic Methods for Partial Differential Equations. Proceedings, 1989. VII, 252 pages. 1990.

Vol. 1451: L. Alvarez-Gaumé, E. Arbarello, C. De Concini, N.J. Hitchin, Global Geometry and Mathematical Physics. Montecatini Terme 1988. Seminar. Editors: M. Francaviglia, F. Gherardelli. IX, 197 pages. 1990.

Vol. 1452: E. Hlawka, R.F. Tichy (Eds.), Number-Theoretic Analysis. Seminar, 1988–89. V, 220 pages. 1990.

Vol. 1453: Yu.G. Borisovich, Yu.E. Gliklikh (Eds.), Global Analysis – Studies and Applications IV. V, 320 pages. 1990.

Vol. 1454: F. Baldassari, S. Bosch, B. Dwork (Eds.), p-adic Analysis. Proceedings, 1989. V, 382 pages. 1990.

Vol. 1455: J.-P. Françoise, R. Roussarie (Eds.), Bifurcations of Planar Vector Fields. Proceedings, 1989. VI, 396 pages. 1990.

Vol. 1456: L.G. Kovács (Ed.), Groups – Canberra 1989. Proceedings. XII, 198 pages. 1990.

Vol. 1457: O. Axelsson, L.Yu. Kolotilina (Eds.), Preconditioned Conjugate Gradient Methods. Proceedings, 1989. V, 196 pages. 1990.

Vol. 1458: R. Schaaf, Global Solution Branches of Two Point Boundary Value Problems. XIX, 141 pages. 1990.

Vol. 1459: D. Tiba, Optimal Control of Nonsmooth Distributed Parameter Systems. VII, 159 pages. 1990.

Vol. 1460: G. Toscani, V. Boffi, S. Rionero (Eds.), Mathematical Aspects of Fluid Plasma Dynamics. Proceedings, 1988. V, 221 pages. 1991.

Vol. 1461: R. Gorenflo, S. Vessella, Abel Integral Equations. VII, 215 pages. 1991.

Vol. 1462: D. Mond, J. Montaldi (Eds.), Singularity Theory and its Applications. Warwick 1989, Part I. VIII, 405 pages. 1991.

Vol. 1463: R. Roberts, I. Stewart (Eds.), Singularity Theory and its Applications. Warwick 1989, Part II. VIII, 322 pages. 1991.

Vol. 1464: D. L. Burkholder, E. Pardoux, A. Sznitman, Ecole d'Eté de Probabilités de Saint- Flour XIX-1989. Editor: P. L. Hennequin. VI, 256 pages. 1991.

Vol. 1465: G. David, Wavelets and Singular Integrals on Curves and Surfaces. X, 107 pages. 1991.

Vol. 1466: W. Banaszczyk, Additive Subgroups of Topological Vector Spaces. VII, 178 pages. 1991.

Vol. 1467: W. M. Schmidt, Diophantine Approximations and Diophantine Equations. VIII, 217 pages. 1991.

Vol. 1468: J. Noguchi, T. Ohsawa (Eds.), Prospects in Complex Geometry. Proceedings, 1989. VII, 421 pages. 1991.

Vol. 1469: J. Lindenstrauss, V. D. Milman (Eds.), Geometric Aspects of Functional Analysis. Seminar 1989-90. XI, 191 pages. 1991.

Vol. 1470: E. Odell, H. Rosenthal (Eds.), Functional Analysis. Proceedings, 1987-89. VII, 199 pages. 1991.

Vol. 1471: A. A. Panchishkin, Non-Archimedean L-Functions of Siegel and Hilbert Modular Forms. VII, 157 pages. 1991.

Vol. 1472: T. T. Nielsen, Bose Algebras: The Complex and Real Wave Representations. V, 132 pages. 1991.

Vol. 1473: Y. Hino, S. Murakami, T. Naito, Functional Differential Equations with Infinite Delay. X, 317 pages. 1991.

Vol. 1474: S. Jackowski, B. Oliver, K. Pawałowski (Eds.), Algebraic Topology, Poznań 1989. Proceedings. VIII, 397 pages. 1991.

Vol. 1475: S. Busenberg, M. Martelli (Eds.), Delay Differential Equations and Dynamical Systems. Proceedings, 1990. VIII, 249 pages. 1991.

Vol. 1476: M. Bekkali, Topics in Set Theory. VII, 120 pages. 1991.

Vol. 1477: R. Jajte, Strong Limit Theorems in Noncommutative L_2-Spaces. X, 113 pages. 1991.

Vol. 1478: M.-P. Malliavin (Ed.), Topics in Invariant Theory. Seminar 1989-1990. VI, 272 pages. 1991.

Vol. 1479: S. Bloch, I. Dolgachev, W. Fulton (Eds.), Algebraic Geometry. Proceedings, 1989. VII, 300 pages. 1991.

Vol. 1480: F. Dumortier, R. Roussarie, J. Sotomayor, H. Żoładek, Bifurcations of Planar Vector Fields: Nilpotent Singularities and Abelian Integrals. VIII, 226 pages. 1991.

Vol. 1481: D. Ferus, U. Pinkall, U. Simon, B. Wegner (Eds.), Global Differential Geometry and Global Analysis. Proceedings, 1991. VIII, 283 pages. 1991.

Vol. 1482: J. Chabrowski, The Dirichlet Problem with L^2-Boundary Data for Elliptic Linear Equations. VI, 173 pages. 1991.

Vol. 1483: E. Reithmeier, Periodic Solutions of Nonlinear Dynamical Systems. VI, 171 pages. 1991.

Vol. 1484: H. Delfs, Homology of Locally Semialgebraic Spaces. IX, 136 pages. 1991.

Vol. 1485: J. Azéma, P. A. Meyer, M. Yor (Eds.), Séminaire de Probabilités XXV. VIII, 440 pages. 1991.

Vol. 1486: L. Arnold, H. Crauel, J.-P. Eckmann (Eds.), Lyapunov Exponents. Proceedings, 1990. VIII, 365 pages. 1991.

Vol. 1487: E. Freitag, Singular Modular Forms and Theta Relations. VI, 172 pages. 1991.

Vol. 1488: A. Carboni, M. C. Pedicchio, G. Rosolini (Eds.), Category Theory. Proceedings, 1990. VII, 494 pages. 1991.

Vol. 1489: A. Mielke, Hamiltonian and Lagrangian Flows on Center Manifolds. X, 140 pages. 1991.

Vol. 1490: K. Metsch, Linear Spaces with Few Lines. XIII, 196 pages. 1991.

Vol. 1491: E. Lluis-Puebla, J.-L. Loday, H. Gillet, C. Soulé, V. Snaith, Higher Algebraic K-Theory: an overview. IX, 164 pages. 1992.

Vol. 1492: K. R. Wicks, Fractals and Hyperspaces. VIII, 168 pages. 1991.

Vol. 1493: E. Benôt (Ed.), Dynamic Bifurcations. Proceedings, Luminy 1990. VII, 219 pages. 1991.

Vol. 1494: M.-T. Cheng, X.-W. Zhou, D.-G. Deng (Eds.), Harmonic Analysis. Proceedings, 1988. IX, 226 pages. 1991.

Vol. 1495: J. M. Bony, G. Grubb, L. Hörmander, H. Komatsu, J. Sjöstrand, Microlocal Analysis and Applications. Montecatini Terme, 1989. Editors: L. Cattabriga, L. Rodino. VII, 349 pages. 1991.

Vol. 1496: C. Foias, B. Francis, J. W. Helton, H. Kwakernaak, J. B. Pearson, H_∞-Control Theory. Como, 1990. Editors: E. Mosca, L. Pandolfi. VII, 336 pages. 1991.

Vol. 1497: G. T. Herman, A. K. Louis, F. Natterer (Eds.), Mathematical Methods in Tomography. Proceedings 1990. X, 268 pages. 1991.

Vol. 1498: R. Lang, Spectral Theory of Random Schrödinger Operators. X, 125 pages. 1991.

Vol. 1499: K. Taira, Boundary Value Problems and Markov Processes. IX, 132 pages. 1991.

Vol. 1500: J. P. Serre, Lie Algebras and Lie Groups. VII, 168 pages. 1992.